Non-Invasive Prenatal Screening of Rare Fetal Genetic Diseases

Non-Invasive Prenatal Screening of Rare Fetal Genetic Diseases

Edited by

Riyaz Ahmad Rather

Department of Bioinformatics, Faculty of Engineering & Technology, Marwadi University, Rajkot, Gujarat, India

ELSEVIER

ACADEMIC PRESS
An imprint of Elsevier

Academic Press is an imprint of Elsevier
125 London Wall, London EC2Y 5AS, United Kingdom
50 Hampshire Street, 5th Floor, Cambridge, MA 02139, United States

ISBN: 978-0-443-27660-6

For information on all Academic Press publications visit our website at https://www.elsevier.com/books-and-journals

Publisher: Megan Ball
Acquisitions Editor: Ali Afzal-Khan
Editorial Project Manager: Rahul Basak
Production Project Manager: Rajeswari R
Cover Designer: Mark Rogers

Typeset by TNQ Tech

Contents

Contributors

Nadir Aman
Utage Child Development Centre, Hyderabad, India; JSS Medical College, Mysuru, Karnataka, India

Onaiza Ansari
Department of Biotechnology, Jamia Hamdard, New Delhi, India

Asma Bashir
General Requirement Department, Fatima College of Health Sciences, United Arab Emirates

Insha Bashir
Department of Biochemistry, GMC, Srinagar, University of Kashmir, J&K, India

Yasir Bashir
Advanced Centre of Human Genetics, Sher-i-Kashmir Institute of Medical Sciences, Srinagar, Jammu and Kashmir, India

Bekele Chakiso
Department of Biology, College of Natural and Computational Science, Wachemo University, Hossana, Ethiopia

Aayushi Gupta
Redcliffe Lab, Noida, Uttar Pradesh, India

Syed Asra Kirmani
Advanced Centre of Human Genetics, Sher-i-Kashmir Institute of Medical Sciences, Srinagar, Jammu and Kashmir, India

Tahir Mohiuddin Malla
Advanced Centre of Human Genetics, Sher-i-Kashmir Institute of Medical Sciences, Srinagar, Jammu and Kashmir, India

Anam Mursaleen
Department of Biotechnology, Jamia Hamdard, New Delhi, India

Aishwarya Panapil
Stand Lifescience Pvt. Ltd., Bangalore, India

Charu Rajpal
Department of Biotechnology, Manav Rachna International Institute of Research and Studies, Faridabad, Haryana, India

Md. Rashid Alam
Department of Biotechnology, Jamia Hamdard, New Delhi, India

Riyaz Ahmad Rather
Department of Bioinformatics, Faculty of Engineering & Technology, Marwadi University, Rajkot, Gujarat, India

Syeda Rushna
Department of Biotechnology, Jamia Hamdard, New Delhi, India

Mohd Shariq
GITAM School of Science, Department of Life Sciences, GITAM University Hyderabad, Hyderabad, Telangana, India

Javaid Ahmad Sheikh
Department of Biotechnology, Jamia Hamdard, New Delhi, India

Wasia Showkat
Advanced Centre of Human Genetics, Sher-i-Kashmir Institute of Medical Sciences, Srinagar, Jammu and Kashmir, India

Vijayalakshmi Srinivasan
MCTM Chidambram Chettyar International School, Chennai, Tamil Nadu, India

Prashant Utage
Utage Child Development Centre, Hyderabad, India

Jaffar Abass Wani
Advanced Centre of Human Genetics, Sher-i-Kashmir Institute of Medical Sciences, Srinagar, Jammu and Kashmir, India

Mahrukh Hameed Zargar
Advanced Centre of Human Genetics, Sher-i-Kashmir Institute of Medical Sciences, Srinagar, Jammu and Kashmir, India

CHAPTER

Introduction to rare genetic diseases in fetal development

1

Asma Bashir

General Requirement Department, Fatima College of Health Sciences, United Arab Emirates

1. Introduction

Normal human development from zygote to fetus is a precisely coordinated process controlled by genetic and epigenetic mechanisms. Interruptions in these mechanisms due to genetic abnormalities can have profound consequences, ranging from fetal mortality to severe congenital anomalies. Multiple studies have shown that a large proportion of spontaneous miscarriages and stillbirths are attributable to chromosomal or genetic disorders (van den Berg et al., 2012; Dolanc Merc et al., 2023). Studies also indicate that over half of first-trimester miscarriages carry chromosomal abnormalities, underlining the critical role of genomic integrity in embryonic viability (Brooks et al., 2022; Vanneste et al., 2009). In the population of couples with recurrent pregnancy loss, genetic factors such as chromosomal translocations in one parent are significantly higher (Jaslow et al., 2010). Genetic disorders are also a leading cause of neonatal morbidity and mortality, with an estimated 20% of infant deaths in developed countries attributable to congenital anomalies, many of which have a genetic basis (Mathews & Driscoll, 2017). Advances in molecular genetics over the past decade have vastly improved our ability to identify and understand prenatal genetic disorders. Genomic technologies such as high-throughput sequencing now allow detection of neonatal and prenatal abnormalities with higher sensitivity, ranging from whole or missing chromosomes to single-nucleotide mutations. Noninvasive prenatal testing (NIPT) has made screening pregnancies for common aneuploidies with exceptional accuracy and safety (Minear et al., 2015; Shaw et al., 2020). However, the genetic implications on fetal development are vast and complex and range from sequence mutations, chromosomal and epigenetic modifications, including even gene-environment interactions. In this review, we are discussing developmental and clinical manifestations of the spectrum of rare genetic disorders impacting fetal development, including chromosomal, monogenic, imprinting-related, and mitochondrial disorders and highlight some molecular and epigenetic mechanisms. We will also briefly discuss genomic screening and diagnosis technologies and their implications in diagnosing genetic conditions prenatally.

Non-Invasive Prenatal Screening of Rare Fetal Genetic Diseases. https://doi.org/10.1016/B978-0-443-27660-6.00007-9

2. Rare genetic disorders and fetal development

Rare diseases is defined different in different countries, with Europe classifying, when it affects no more than one person in every 2000 (Abozaid et al., 2025) while in US, a rare disease impacts fewer than 200,000 individuals nationwide, while in Japan, the threshold is set at fewer than 50,000 people, or roughly one in 2500 individuals (Hayashi & Umeda, 2008). Rare genetic diseases collectively affect approximately 1 in 20 individuals worldwide, presenting with diverse and severe phenotypes including congenital abnormalities and neurodevelopmental disorders (Dawood et al., 2024). Genetic disorders that affect fetal development can be broadly classified into several overlapping categories: chromosomal abnormalities, monogenic (single-gene) disorders, genomic imprinting disorders, and mitochondrial disorders. Each category encompasses a range of conditions that can influence embryogenesis and fetal growth in distinct ways (Fig. 1.1).

2.1 Chromosomal anomalies

Chromosomal abnormalities are a common and significant cause of human pregnancy loss and fetal issues (Hyde & Schust, 2015). These abnormalities include aneuploidies (abnormal number of chromosomes), polyploidies, and chromosomal rearrangements. The aneuploidy in humans can be of two types: (a) *homogenous aneuploidy*, where all cells of an embryo contain the same number (incorrect) of chromosomes, and (b) *Mosaic aneuploidy*, where cells of one embryo contain mixtures of euploid and aneuploid cells. Homogeneous aneuploidies originate during gamete formation because of an error in chromosome segregation, resulting in the gain or loss of a chromosome. The most frequently observed homogenous aneuploidies in liveborn infants include trisomies (3 copies) of chromosomes 21, 18, and 13 (Down syndrome, Edwards syndrome, and Patau syndrome, respectively) (Munne et al., 1998). Trisomy 21, for example, occurs roughly 1 in 700 live births and, while compatible with survival, leads to characteristic dysmorphic features and neurodevelopmental impairment. In contrast, trisomies 18 and 13 are more severe, with high rates of in utero demise and neonatal mortality. Aneuploidy is the most common reason for abortions, accounting for at least 50% of first-trimester miscarriages (Brooks et al., 2022). Polyploidy, such as triploidy, invariably result in pregnancy loss or early neonatal death. Notably, triploidy can arise from fertilization errors and is associated with abnormal placental development, severe growth restriction, and structural defects in the fetus (Masset et al., 2021).

Errors in the very first zygotic division result in embryos that are mosaics—for example, one blastomere inherits both copies of each parental genome (effectively diploid biparental), while another blastomere gets two paternal sets (androgenetic diploid) and yet another gets two maternal sets (gynogenetic diploid), or various haploid complements (De Coster et al., 2022). Masset and colleagues proposed this "heterogenic division" model that provides a single mechanism for the origin of mixoploidy and chimerism: where a single embryo cleaves into distinct genetic

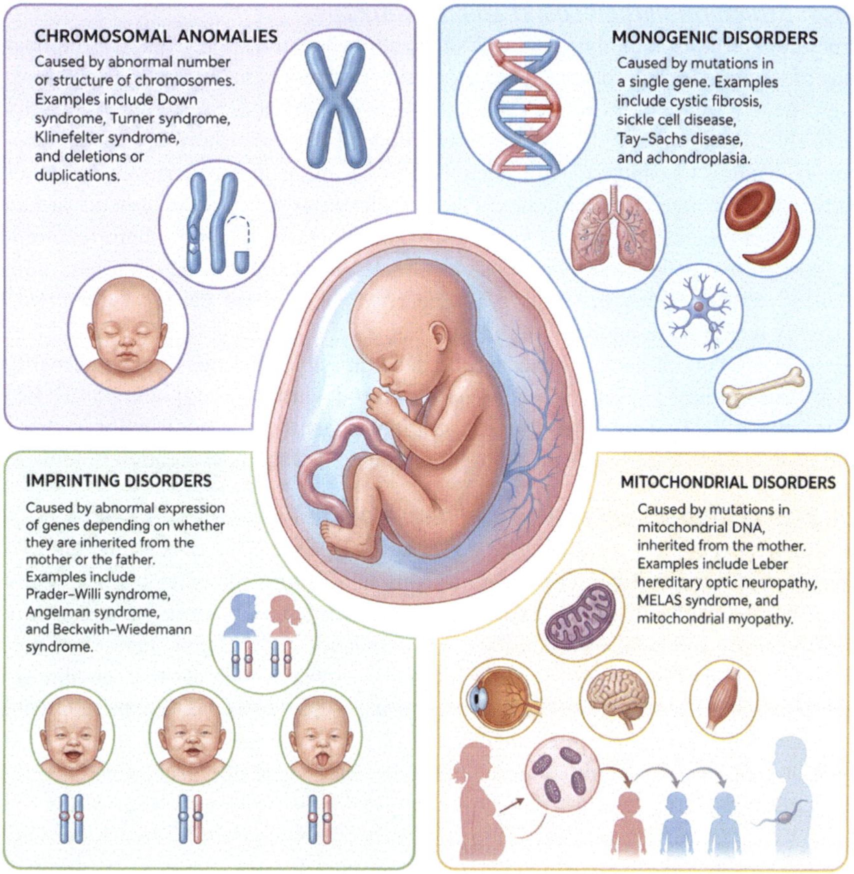

FIGURE 1.1

Genetic disorders. Main types include chromosomal anomalies, monogenic disorders, imprinting disorders and mitochondrial disorders.

lineages due to segregation of entire parental genome sets rather than the normal mixing (Masset et al., 2021). At the molecular level, this could be due to the formation of multiple spindle poles in the zygote or a failure of the male and female pronuclei to merge before division, allowing each pronucleus to segregate separately (De Coster et al., 2022). This results in a patchwork embryo where various cells have different ploidies or parental contributions, which can disrupt the proper signaling essential between cells during morphogenesis. Surviving infants with diploid-triploid mosaicism have been reported and often exhibit dysmorphic features, growth abnormalities, and organ malformations; intriguingly, their phenotypes can sometimes resemble those of imprinting disorders due to the presence of an extra set of paternal or maternal chromosomes in a subset of cells (Destouni et al., 2016).

2.1.1 Structural chromosomal abnormalities

Structural chromosomal abnormalities such as translocations, deletions, or duplications of segments also impact fetal development. Carriers of balanced translocations, where genetic material is rearranged but not lost or gained, are typically healthy themselves but are at risk of producing unbalanced gametes (more or less chromosomes), leading to embryos with aneuploidies resulting in miscarriages or babies with multiple congenital anomalies. Parental chromosomal rearrangements are an important etiologic factor in pregnancy loss, as shown by a study where recurrent pregnancy loss patients showed chromosomal abnormality in 7.6% cases, with balanced translocations accounting for nearly half of these cases (Turki et al., 2016). A recent meta-analysis found that fetuses diagnosed with congenital heart disease (CHD) on ultrasound have significant identifiable chromosomal abnormality (23%), aneuploidy (19%) and pathogenic microdeletions or duplications (~4%) (Wang et al., 2023). Thus, chromosomal anomalies are a major contributor to structural birth defects, thus highlighting the importance of genetic evaluation in fetuses.

2.1.2 Gene dosage imbalance and aneuploid effect on development

Aneuploids experience a global gene dosage imbalance that the developing embryo often cannot tolerate. Each chromosome consists of hundreds of genes; a trisomy (extra copy) results in 150% expression of those genes, whereas a monosomy (missing copy) yields 50% expression. This imbalance disrupts gene regulatory networks and cellular homeostasis. Studies of early embryos indicate that aneuploidy stimulates significant changes in gene expression, proteostasis, and metabolism, often activating stress responses such as the p53 pathway and leading to cell cycle arrest or apoptosis. In preimplantation embryos, time-lapse microscopy and single-cell sequencing have demonstrated how aneuploid cells frequently form during the first few cleavage divisions. For example, it has been observed that chromosomes lag or missegregate during live imaging of bovine and human zygotes, resulting in daughter blastomeres with atypical chromosome counts (Brooks et al., 2022). These errors often lead to failure of cytokinesis or mitotic spindle defects in the embryo at the cleavage stage. In a study by Chavez and colleagues, micronuclei (small extranuclear chromosome-containing bodies) and chromatin bridges between blastomeres were observed, which are indicators of genomic instability in these early divisions (Chavez et al., 2012). These irregularities can result in mosaicism, where some embryonic cells are euploid and others aneuploid. The aneuploid cells in an embryo can have various outcomes—they may undergo apoptosis, preferentially be allocated to extraembryonic tissues, or, in certain situations, continue to proliferate, which results in fetal mosaicism (Singla et al., 2020). A typical example is confined placental mosaicism (CPM), where the placenta has an abnormal karyotype (often trisomic) that is not present in the fetus. CPM results from a postzygotic mutation in the trophoblast lineage or a trisomic conception that "corrects" itself in the inner cell mass. In CPM, the placenta might have a mix of normal and aneuploid cells at the molecular level. Although the fetus contains normal chromosomes, the aneuploid cell population in the placenta can result in placental dysfunction. CPM has been confirmed by detailed molecular cytogenetic studies in cases where unexplained fetal growth restriction (FGR) was observed (Nagamatsu et al., 2014).

Specific molecular players in maintaining chromosome segregation fidelity in the embryo are being identified by mechanistic studies in model systems and human embryos. In one such study, researchers experimentally depleted BUB1B (mitotic checkpoint protein-important for preventing premature anaphase) in early cleavage-stage embryos, which resulted in chromosomal missegregations, chaotic aneuploidies, and deregulation of the kinase signaling network that regulates mitosis (Brooks et al., 2022). This experiment demonstrated that embryonic cells depend upon some of the same checkpoint mechanisms as somatic cells. However, if there is an impairment due to maternal age related degradation of proteins or mutations, the embryo is at high risk of aneuploidy before even activating its own genome. Also, in chimerism, a mosaic embryo where various cells have different ploidy or parental contributions can disrupt the proper signaling essential between cells during morphogenesis.

Most of the aneuploidies are usually fatal, and only a few aneuploidies (such as trisomy 21) sometimes survive. The most reasonable explanation being certain chromosomes contain genes intolerant to dosage change; for example, chromosome 1 has many important genes, and no trisomy of chromosome 1 progresses far in development. In contrast, chromosome 21 has the lowest number of genes, which might partly explain the viability of trisomy 21. Even though the viable aneuploidies lead to many downstream molecular disturbances, the presence of mechanisms such as dosage compensation (X-chromosome inactivation) can reduce effects for certain aneuploidies (e.g., extra X chromosomes). Transcriptome analyses of the fetuses with Down syndrome not only exhibit dysregulation of chromosome 21 genes but also secondary effects on other pathways, such as mitochondrial function and interferon signaling, due to the attempt of cells to adapt to the initial imbalance. The few that survive to term in Edwards and Patau syndromes (trisomies 18 and 13) have severe structural defects; as research indicates these trisomies disrupt developmental gene expression programs (such as HOX gene patterns and cardiac transcription factors) early in organogenesis, resulting in characteristic malformations. Therefore, gene dosage imbalances result in cascades of molecular dysfunction—from cell cycle checkpoints and metabolic stress to modified developmental gene networks—leading to developmental failure or anomalies.

2.2 Monogenic disorders

Monogenic disorders (with errors in a single gene) account for 80% of rare diseases in humans and profoundly affect fetal development (Condo, 2022). With over 4000 known monogenic diseases, many have prenatal onsets, particularly those disrupting fundamental developmental pathways or metabolic functions. For example, certain skeletal dysplasia caused by dominant mutations lead to extreme shortening of limbs and thoracic insufficiency, often resulting in perinatal lethality. Similarly, recessive metabolic disorders such as lysosomal storage diseases with enzymatic dysfunction leading to substrate accumulation can present with nonimmune hydrops fetalis (generalized edema of the fetus) and organomegaly during pregnancy (Platt et al., 2018). Monogenic disorders often show abnormalities in prenatal ultrasound, such as in osteogenesis imperfecta (a brittle bone disease due to collagen gene mutations), multiple fetal fractures and bone deformities are often seen on ultrasound. Autosomal recessive polycystic kidney disease shows enlarged echogenic kidneys and oligohydramnios (low amniotic fluid) in utero. However, the identification of a specific monogenic condition requires molecular testing, as many diseases have overlapping prenatal phenotypes. Certain populations with high rates of consanguinity, such as the Gulf, experience a heavier burden of autosomal recessive disorders in fetuses and newborns, due to increased homozygosity for deleterious mutations (Chekroun et al., 2025). The Gulf countries harbor numerous distinct founder mutations and a high prevalence of recessive diseases, which results in more children with genetic disorders and fetal demise or neonatal intensive care admissions (Munné, 2006; Springett et al., 2015).

A gene's function is responsible for the molecular dysfunction seen in single-gene mutations with many monogenic disorders that occur in utero, involving loss of function in genes important for organ development. For example, mutations in collagen type I genes (COL1A1 or COL1A2) in osteogenesis imperfecta lead to defective collagen fibers. In a fetus, this molecular defect results in brittle bones that fracture under normal intrauterine pressures, and damaged bone mineralization is observed as skeletal deformities on ultrasound. Similarly, a mutation in the cardiac transcription factor gene, NKX2-5, can lead to structural heart defects such as septal defects and conduction anomalies by disrupting the gene regulatory networks that form the developing heart tube. Inborn errors of metabolism generally occur when the fetus transitions to postnatal life, for example, phenylketonuria typically occurs after birth when certain substrates accumulate from feeding. Although some metabolic disorders do occur prenatally, such as Mucopolysaccharidosis type VII (Sly syndrome), which leads to hydrops fetalis due to lysosomal storage of glycosaminoglycans, and X-linked ornithine transcarbamylase (OTC) deficiency, which has caused prenatal onset hyperammonaemia and cerebral edema in male fetuses that ultimately resulted in miscarriage. The molecular mechanism in metabolic disorders is substrate accumulation or product deficiency that disturbs cellular homeostasis. In OTC deficiency, the urea cycle is blocked, leading to ammonia accumulation, which is neurotoxic, resulting in a severely affected fetus that can succumb to a metabolic crisis in utero.

A notable subset of monogenic disorders that affect fetal development involves genes involved in developmental signaling pathways (e.g., fibroblast growth factor, hedgehog, or retinoic acid pathways). Mutations in these pathways can lead to malformations by changing cell fate decisions and tissue patterning. For instance, mutations in DHCR7 cause Smith-Lemli-Opitz syndrome (SLOS), a recessive metabolic disorder of cholesterol synthesis that results in low cholesterol and accumulation of toxic precursors. As cholesterol is essential for Sonic hedgehog (SHH) signaling, if the SHH pathway is disrupted, SLOS causes a broad range of malformations, including holoprosencephaly (forebrain patterning defect) and limb abnormalities. Thus, a single enzymatic defect is responsible for a cascade of developmental signaling failures. On a molecular level, this demonstrates the principle of pleiotropy, where one gene influencing multiple organ systems is common in developmental genetics because key signaling molecules are reused in multiple contexts (e.g., SHH in the brain, limb, and face). Hence, a mutation in one component can result in a syndrome of defects.

2.3 Imprinting disorders

In humans, many genes are subject to genomic imprinting, meaning that only one parental allele is actively expressed while the other is silenced by epigenetic markers established in the germline (Monk et al., 2019). Genomic imprinting is as important as the DNA sequence, and an imbalance caused by disruptions of the normal parent-of-origin-specific expression of genes can cause disorders. Imprinting disorders involve epigenetic error, such as abnormal DNA methylation patterns at imprinting control regions or uniparental disomy (UPD), wherein both copies of a chromosome or part of a chromosome come from one parent, effectively leading to double dosage of one imprint and loss of the other. Imprinting disorders can significantly affect fetal development and the pregnancy course, and as persist in all or most somatic cells, leading to lifelong phenotypic effects (Monk et al., 2019; Wojcik et al., 2018). Imprinting disorders include Beckwith–Wiedemann syndrome (BWS) and Silver–Russell syndrome (SRS), both linked to imprinting errors on chromosome 11p15 involving genes that regulate growth. BWS most often results from loss of maternal methylation (or paternal UPD) on this chromosome and is characterized by fetal and neonatal overgrowth, organomegaly, and higher risk of embryonal tumors. SRS often involves loss of paternal methylation (or maternal UPD) and has severe intrauterine growth restriction and short stature (Monk et al., 2019). Other common imprinting disorders, such as Prader–Willi syndrome (PWS) and Angelman syndrome (AS), caused by deletions or UPD of chromosome 15q11-13, do not show significant structural anomalies on prenatal imaging. However, it manifests mainly in the postnatal period, with PWS showing hypotonia and feeding difficulty in infancy and AS with severe developmental delays and seizures.

Genomic balance between maternal and paternal contributions in genome imprinting is necessary for fetal development. A rare ovarian teratoma (i.e., benign tumors containing differentiated fetal tissues) can arise from unfertilized oocytes undergoing parthenogenetic division, thus necessitating the requirement of a proper biparental genome for normal embryogenesis. Imbalance can also occur in the form of androgenetic–biparental mosaicism, wherein the placenta contains a mixture of normal and androgenetic (paternal-only) cell line leading to placental mesenchymal dysplasia and severe fetal growth restriction (Lee et al., 2020; Masset et al., 2021).

2.3.1 Epigenetic regulation and imprinting in fetal development

Epigenetic modifications are chemical marks on DNA and histones that affect gene expression without changing the DNA sequence. It plays a significant role in development, where during gametogenesis and early embryogenesis, the genome undergoes extensive epigenetic reprogramming, including the establishment (in gametes) and maintenance (in the embryo) of imprinting marks at certain loci. These imprinted loci are controlled by DNA methylation, specific to alleles, resulting in expression that depends on parent-of-origin. Disturbances of these epigenetic modifications can disrupt normal developmental processes in a subtle or specific manner. For example, the imprinting control region (ICR) on chromosome 11p15.5 controls a group of growth-related genes such as IGF2 and H19. In normal development, the maternal allele at ICR is methylated, resulting in inhibition of a growth promoter, IGF2 and expression of a growth suppressor, H19, while the paternal allele is unmethylated, allowing IGF2 expression. If there is any imbalance, it will lead to growth restriction by a loss of methylation on the maternal ICR, where both alleles will behave as maternal (with IGF2 off and H19 on), as seen in Silver–Russell syndrome. Conversely, if both alleles are methylated (paternal pattern), IGF2 is double-expressed, and H19 is inhibited, which will lead to overgrowth as seen in Beckwith–Wiedemann syndrome (BWS) (Beygo et al., 2013; Cerrato et al., 2008). Alterations in methylation, at a molecular level, can occur due to mutations in the DNA sequence of elements that regulate imprinting, or they can be primary epimutations, which are errors in either the establishment or maintenance of methylation with no underlying DNA mutation (Cerrato et al., 2008). Primary epimutations may be triggered by environmental factors or occur randomly. Research has proven that assisted reproductive technologies (ART), such as in vitro fertilization and manipulations of gametes/embryos, are related to a slightly increased occurrence of certain imprinting disorders such as BWS (Cortessis et al., 2018). One such hypothesis is that ART procedures occur during important processes of epigenetic reprogramming and might hinder the proper establishment of imprint marks, through culture conditions or ovary stimulation hormones. Similarly, other environmental factors (maternal nutrition, toxin exposures) are being studied for their association with imprinting errors, though authentic data in humans remain limited. It is known that once an imprinting error is present in the early embryo, all descendant cells will carry that epigenetic state, leading to consistent mis-expression of the affected imprinted gene in fetal tissues (Monk et al., 2019). This is the reason that a primary epimutation causing loss of maternal methylation at the PWS/Angelman locus will result in a child with Angelman syndrome (if the normally active maternal allele of UBE3A is silenced by the epimutation), even though the DNA sequence is normal. The developmental consequence of imprinting disorders involves disordered growth and metabolism. Many imprinted genes are fetal growth regulators, aligning with the "parental conflict" theory, which suggests paternal genes promote growth while maternal genes inhibit it. According to this, conditions such as BWS (paternal domination) show macrosomia and organ hypertrophy, whereas SRS (maternal domination or paternal loss) show failure of growth. These growth effects can be associated with pathways such as insulin-like growth factor signaling at the cellular level. Increased cell proliferation and enlarged size of organs such as liver, tongue, and adrenal glands are found in BWS due to biallelic expression of IGF2 in the fetus and placenta, which also leads to embryonal tumor formation (e.g., Wilms tumor) in childhood as a result of excessive growth signals (Monk et al., 2019). On the other hand, in PWS (loss of paternally expressed genes on 15q11), hypothalamic dysfunction occurs, which leads to neonatal hypotonia and later hyperphagia, where one of the affected genes, MAGEL2, is associated with neurodevelopmental pathways. This demonstrates how a defect in imprinting can particularly target neural circuits that control appetite and muscle tone. It is interesting to note that if an imprinting or epigenetic error occurs mosaically and not in every cell, the phenotype can be less severe or variable. For instance, mosaic uniparental disomy might be responsible for segmental or milder forms of disorders. Documented cases of mosaic BWS have indicated that only a few tissues (arising from one early cell lineage) have the methylation defect, leading to asymmetric or patchy overgrowth (a phenomenon known as hemihypertrophy). This emphasizes that if the error occurs after the first few cell divisions, epigenetic regulation can vary even between cells of the same individual.

Besides imprinting, other epigenetic processes are significant in development, such as X-chromosome inactivation (XCI) in females. Abnormalities in XCI (such as skewed inactivation of the X carrying a critical gene, or failure to inactivate an extra X) can result in developmental issues, even though it is not classically termed as an "imprinting disorder." For example, in Turner syndrome (45, X), there is no second X to inactivate, and the single X may overexpress some genes that normally escape XCI, contributing to the phenotype. In trisomy X (47, XXX), although the two Xs are largely inactivated, an overdose of the genes that escape inactivation can result in tall stature and mild cognitive effects. These examples demonstrate how epigenetic dosage compensation mechanisms align with chromosomal abnormalities.

2.4 Mitochondrial disorders

Mitochondrial diseases form another category of rare genetic disorders affecting fetal development with unique inheritance patterns and cellular pathology. The mitochondrial genome (mtDNA) is inherited exclusively from the mother and exists in multiple copies per cell. Pathogenic mutations in mtDNA can lead to a state where a mixture of mutant and wild-type mtDNA coresides in a cell, called as heteroplasmy (Gorman et al., 2016).

Mitochondrial dysfunction is often determined by the amount of mutant mtDNA, as there is often a threshold effect where clinical symptoms occur only if the fraction of mutant genomes in a tissue surpasses a specific level. During embryonic development, the random distribution of mitochondria to daughter cells can lead to differences in the mitochondrial content across different tissues of the fetus, which in turn contributes to the phenotypic variability seen in mitochondrial disorders

(Stewart & Chinnery, 2015). Mutations in nuclear genes encoding mitochondrial proteins result in some mitochondrial disorders; even though these genes follow Mendelian inheritance (often autosomal recessive), they disrupt mitochondrial function. Because mitochondria are essential for energy production, high-energy tissues such as the brain, heart, skeletal muscle, and placenta are particularly susceptible to mitochondrial defects. Severe mitochondrial disorders can occur in utero or in the neonatal period with characteristics such as cardiomyopathy, growth restriction, or lactic acidosis. For example, mitochondrial DNA depletion syndromes (due to nuclear gene defects causing reduced mtDNA copy number) generally lead to prenatal onset of growth failure and liver dysfunction. Mutations in either mtDNA or nuclear genes also lead to Leigh syndrome, a severe neurodegenerative disorder. The clinical and genetic heterogeneity of mitochondrial disorders also complicates prenatal prediction; a fetus that inherits an mtDNA mutation may be asymptomatic at birth or can have a severe neonatal illness, considering if the mutant load in critical tissues is low or high, respectively. Furthermore, mitochondrial genetics allows unique phenomena (dramatic shifts in heteroplasmy across generations) and maternal mosaicism, which complicates genetic counseling. Nonetheless, mitochondrial disorders are an integral part of the landscape of genetic fetal disorders. Collectively, mitochondrial diseases rank among the most common inherited metabolic disorders, even though individually they are rare (incidences ranging from 1 in 10,000 to 1 in 100,000). Research into mitochondrial pathophysiology has shown that these diseases are not entirely about energy deficits but can affect development due to stimulation of stress pathways, altered apoptosis, and changes in metabolic signaling (Wen et al., 2025). For instance, mitochondrial mutations impair oxidative phosphorylation in placental cells, which leads to placental insufficiency and fetal growth restriction in certain pregnancies, even though the fetus's own tissues are not yet severely affected. Therefore, understanding the spectrum of genetic disturbances is the key step in developing effective strategies for their detection, management, and prevention. The following sections will delve into how these genetic abnormalities affect molecular mechanisms, how we can identify them before birth, and what measures can be taken to reduce their impact.

2.4.1 Mitochondrial dysfunction pathways affecting fetal development

There is more to the molecular pathology of mitochondrial disorders than just "less ATP." Mitochondria are involved in many cellular processes, including apoptosis, calcium homeostasis, and the production of reactive oxygen species (ROS). Cells generally activate compensatory and stress response pathways when there is an irregular mitochondrial function due to a genetic mutation. Recent research has demonstrated the role of the mitochondrial integrated stress response (mt-ISR), a program that adjusts nuclear gene expression in response to mitochondrial dysfunction (e.g., upregulating mitochondrial chaperones and antioxidant enzymes) (Wang & Zhang, 2025). In developing tissues, if mitochondrial energy output is insufficient, cells may switch to anaerobic glycolysis (the Warburg effect) to compensate for ATP. Thus, fetuses with mitochondrial disorders sometimes show elevated levels of lactic acid (an indication of anaerobic metabolism). However, some mitochondrial mutations do not dramatically decrease ATP levels due to partial compensation but cause disease through other mechanisms. For example, specific cell types experience oxidative damage and impaired enzyme activity due to a defect in the assembly of mitochondrial iron-sulfur clusters (as in Friedreich's ataxia, a nuclear gene disorder affecting mitochondrial function), even if production of ATP is slightly affected (Rotig et al., 1997). Such milder mitochondrial dysfunction might not kill a cell directly but could interfere with differentiation or organ maturation in fetal development. Also, cells have quality control mechanisms such as mitophagy (selective autophagic removal of damaged mitochondria) and mitochondrial biogenesis signals that try to maintain a healthy mitochondrial population. If these quality controls are disturbed by a genetic defect, the result can be apoptosis or failure of those cells to contribute properly to the developing tissue. For example, in the developing brain, neurons are highly energy-dependent; if a fraction of neurons undergo apoptosis due to overwhelming mitochondrial stress, the brain may develop abnormally large ventricles or cortical thinning. This has been observed in some mitochondrial encephalopathies detectable on fetal MRI.

The potential of intergenomic interaction is also an interesting feature of mitochondrial genetics; mutations in mtDNA and mutations in nuclear DNA encoding mitochondrial proteins can sometimes produce similar traits, but the cell's response may vary. Current research is investigating mitochondrial-nuclear interaction; for example, transcripts encoded in the nucleus can alter to offset mtDNA mutations (upregulating alternative metabolic pathways). While mtDNA mutations may cause a retrograde response due to respiratory chain imbalance, nuclear DNA mutations usually stimulate the unfolded protein response in mitochondria as the defective proteins accumulate. Some mtDNA mutations can be effectively mitigated if a tissue manages to increase its mitochondrial number or if a reparative nuclear genetic variation is present. Although when compensatory mechanisms fail, cells accumulate damage, usually through excess ROS production, leading to damage of lipid, protein, and DNA, further aggravating developmental defects. Therefore, understanding both the mutation as well as how the cells of an embryo deal with mitochondrial stress (through biogenesis, metabolic reprogramming, mitophagy, etc.) is necessary in predicting the severity of a mitochondrial disorder from a molecular perspective.

3. Advances in genomic diagnostics and prenatal screening

Early and accurate detection of genetic disorders in the fetus has been a goal of perinatal medicine for a long time. Over the past decade, there have been dramatic advances in prenatal diagnostic and screening technologies, changing how we identify genetic abnormalities before birth. These advances range from noninvasive screening tests capable of analyzing fetal DNA circulating in maternal blood, to

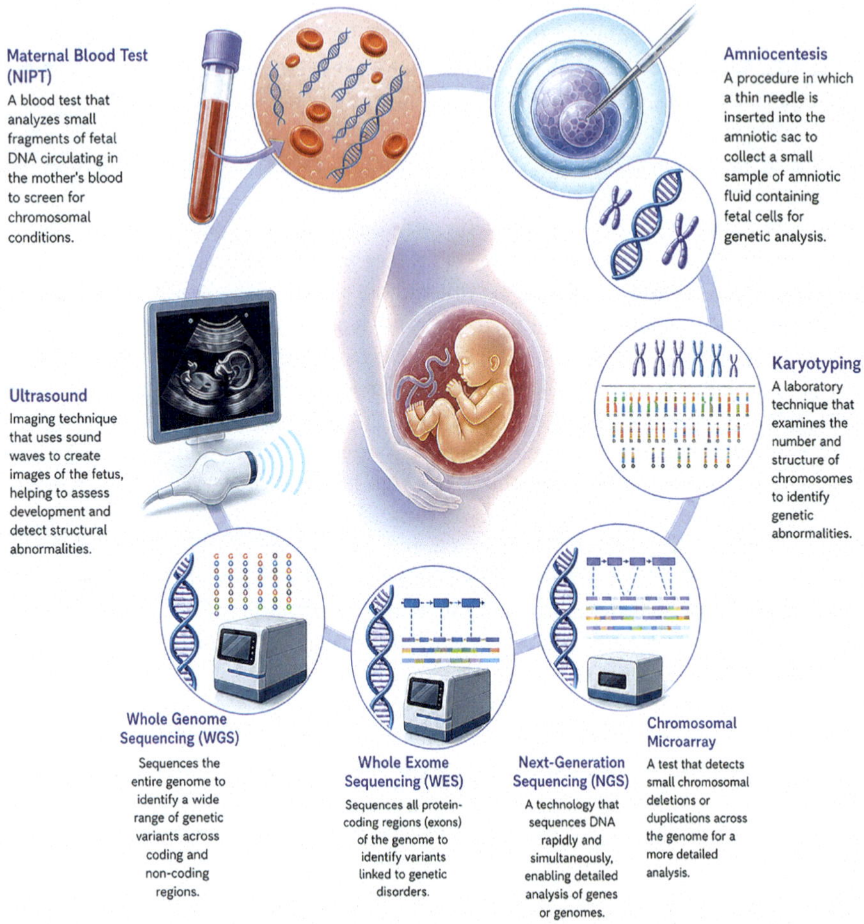

FIGURE 1.2

Prenatal screening techniques. Main techniques include Non-invasive prenatal testing (NIPT), Ultrasound, Non-invasive prenatal diagnosis (NIPD), Chromosomal microarray (CMA), Next generation sequencing (NGS) including whole genome sequencing (WGS), Whole exome sequencing (WES).

high-resolution genomic diagnostics of fetal cells. In this section, we review the advances in prenatal genetics and discuss how these technologies are implemented and are rapidly evolving (Fig. 1.2).

3.1 Noninvasive prenatal testing (NIPT)

DNA from the developing fetus can be detected in maternal plasma and is referred to as cell-free fetal DNA (cffDNA). cffDNA originates primarily from the trophoblasts of the placenta and can be detected as early as 5–7 weeks of gestation (Lo et al., 1997; D'Aversa et al., 2018). By around 15 weeks of gestation, a sufficient fraction of circulating DNA in maternal blood (typically ~5–15%) is fetal-derived, which can be used for genomic analysis of the fetus without an invasive procedure (Palomaki et al., 2011). Use of cffDNA for prenatal genetic testing has been the most transformative innovation, as it avoids the miscarriage risks associated with invasive procedures such as amniocentesis or chorionic villus sampling (CVS) (Alfirevic et al., 2017). This is referred to as noninvasive prenatal testing (NIPT) and has been routinely incorporated clinically for testing aneuploidy since 2008 (Lo et al., 2007). NIPT has quickly been adopted worldwide as a screening tool for common aneuploidies, replacing traditional serum and ultrasound screening and is now available in 60 countries worldwide (Chiu et al., 2011; Ravitsky et al., 2021). NIPT can detect trisomies such as trisomy 21, 18, and 13 with extraordinarily high sensitivity and specificity, leading to increasingly recommendation of it as a primary screening option (Gil et al., 2017; Norton & Wapner, 2015). Current guidelines from obstetric societies such as the American College of Obstetricians and Gynecologists (ACOG) and American College of Medical Genetics and Genomics (ACMG) state that all pregnant patients should be offered or informed about aneuploidy screening or diagnostic testing, regardless of maternal age or baseline risk (Gregg et al., 2016; Rose & Kaimal, 2020). This shift happened as evidence showed efficacy in average-risk cohorts, also and not only in high-risk pregnancies (Christiaens et al., 2021). NIPT has a broadened scope now from trisomies to a range of chromosomal abnormalities, including Turner syndrome and Klinefelter syndrome. Genome-wide NIPT approaches are now tested to detect large copy number variations and microdeletion syndromes (e.g., 22q11.2 deletion/DiGeorge syndrome, 1p36 deletion, etc.) (Xue et al., 2022). This means NIPT could potentially catch a variety of conditions such as Cri-du-chat syndrome, caused by a deletion, which would previously have gone unnoticed until birth or invasive testing.

Although NIPT detection and scope have increased with time, it also introduces new challenges. The positive predictive value (PPV), which represents reliability for determining the positive of NIPT, is much lower for rare conditions than for common trisomy cases. This is true for sex chromosome aneuploidies, where false-positive rates of up to 90% have been reported in low-risk cohorts (Reiss et al., 2017). False-positive NIPT results can occur because of the different genetic makeup of the placenta compared with the fetus, called as confined placental mosaicism (CPM) (Reiss et al., 2017). Other reasons can be maternal chromosomal anomalies such as undiagnosed copy number variations, triple X syndrome, or mosaic Turner's syndrome (Zhang et al., 2020). A vanished twin with aneuploidy or circulating tumor DNA derived from mothers can also give a false signal (Bianchi et al., 2015). Another limitation of NIPT can be false negative results, which are very low (0.01%–0.1%) for common trisomies such as Down syndrome, but not zero. This can be due to low fetal DNA fraction seen in early gestation, high maternal weight or even medications (Kuhlmann-Capek et al., 2019).

These issues have raised ethical and clinical concerns related to NIPT results, as they can cause unnecessary anxiety in mothers. Thus, caution is advised in the use of NIPT beyond common trisomies, and positive results need confirmation with diagnostic testing such as chorionic villus sampling or amniocentesis before clinical decisions (Rose & Kaimal, 2020). Despite limitations, NIPT has indeed reduced the number of invasive tests performed and the number of missed aneuploidies. NIPT has led to raised detection of Down syndrome before 20 weeks, and a drop in invasive procedures by as much as 70%–80% in some regions, with fewer women requiring confirmatory testing. Many countries, such as the United Kingdom, have integrated NIPT into their public health programs for prenatal care, although with different models. In the UK and many European countries, NIPT is available for high-risk mothers, whereas in the US, it's offered universally in the private sector (Jayashankar et al., 2023). Access and insurance coverage in different countries are still evolving, with most providing funding for high-risk mothers.

3.2 Noninvasive prenatal diagnosis (NIPD)

Researchers are now extending cell-free DNA analysis beyond aneuploidy screening to single-gene disorders, often termed noninvasive prenatal diagnosis (NIPD). NIPD for monogenic conditions can be targeted against a known mutation/variants inherited from parents, de novo variants or haplotype dosage analysis. Unlike NIPT, where a confirmatory invasive testing is required for aneuploidy, NIPD for monogenic disease in high-risk pregnancies (family history or suggestive ultrasound findings) can be diagnostic because parental genomes are analyzed in parallel to account for confounding factors (such as CPM). Current clinical applications of NIPD include fetal RhD blood type genotyping for Rh-negative mothers to prevent hemolytic disease of the fetus and newborn (HDFN), and sex determination where ultrasound is ambiguous (Devaney et al., 2011; Legler et al., 2021). NIPD is now being tested out in high-risk patients for many monogenic conditions, such as *FGFR2*/*FGFR3*-related skeletal conditions, Duchenne/Becker muscular dystrophies, cystic fibrosis, and spinal muscular atrophy (Chandler et al., 2020; Chitty et al., 2015; Parks et al., 2016, 2017). For autosomal recessive diseases such as cystic fibrosis and spinal muscular atrophy, relative haplotype dosage (RHDO) is being used, where SNPs are used to trace the at-risk haplotype. NIPD still face multiple challenges main one being low fetal fractions (FF) relative to maternal DNA, requiring extremely sensitive assays and inconclusive results. Other challenges include maternal somatic mosaicisms, consanguinity and recombination events within the target locus leading to inconclusive results (Chandler et al., 2020). The need for techniques such as RHDO also creates a technical and economic barrier.

3.3 Ultrasound and traditional screening

The role of prenatal ultrasound remains indispensable even though genomic tests are growing to the forefront. Ultrasound can detect structural anomalies such as heart defects, spina bifida, brain malformations, and skeletal dysplasia that may be the first clue to an underlying genetic condition. Soft markers for aneuploidy, such as increased nuchal translucency, cardiac echogenic focus, can also be identified. An increased nuchal translucency (NT) measurement often correlates with chromosomal abnormalities and some monogenic syndromes (Nicolaides et al., 1992). Ultrasound is a screening tool, but may create suspicions that can lead to the detection of genetic conditions (such as chromosomal, microdeletions or single-gene disorders). According to ACOG, Patients showing significant fetal anomaly in ultrasound should be offered diagnostic genetic testing (such as chromosomal microarray, exome sequencing) without considering NIPT results (Rose & Kaimal, 2020).

3.4 Chromosomal microarray (CMA)

Chromosomal microarray has become a powerful diagnostic tool in prenatal genetics and can detect aneuploidies and copy number variants (CNVs) (Levy & Wapner, 2018). According to the 2010 consensus, CMA is considered a first-tier clinical diagnostic test for individuals with developmental or intellectual disabilities or congenital abnormalities (Miller et al., 2010). CMA can also indicate a risk of recessive disorders in populations with high consanguinity rates, such as in the Gulf (El Naofal et al., 2023). The integration of CMA has improved the ability to provide specific diagnoses, prognoses, and recurrence risk counseling early and prenatal management. CMA frequently identifies variants of unknown pathogenicity called variants of uncertain significance (VOUS), causing maternal anxiety, for example, Wang et al. showed that about 4% of fetuses with CHD had VOUS (Wang et al., 2023). Over time, as databases such as ClinVar grow, some of these VOUS will get reclassified (often as benign).

3.5 Prenatal exome and genome sequencing

The frontier of prenatal diagnosis is moving toward whole exome sequencing (WES) where the coding regions (exome) are sequenced, and whole genome sequencing (WGS), which involves sequencing the whole genome of the fetus. WGS is feasible but costly, with huge data to interpret compared with WES. WGS/WES are useful for detecting monogenic or rare genetic disorders, which are not identifiable by karyotype or microarray. Rapid sequencing, such as STATseq, is increasingly utilized for newborns with suspected monogenic disorders. Neonates presenting complex phenotypes, including neuromuscular disease, metabolic crises, or cardiovascular malformations, are generally offered such tests (Saunders et al., 2012). Multiple studies such as the PrenatalSEQ study (USA), and the PAGE study (UK) have used WGS/WES to diagnose fetal structural anomalies, and typical yields reported are 8%–10% for fetuses with a single major anomaly and up to 30% for those with multiple anomalies (Lord et al., 2019; Petrovski et al., 2019). Clinical WES has seen to achieve a molecular diagnosis in approximately 36.7% of infants, while WGS has demonstrated yields as high as 50%–70% in research settings (Meng et al., 2017).

The use of parent-infant trios is the general criterion, as it leads to the identification of de novo mutations and identifies benign familial variants. Targeted gene panels are also used to target selected genes for possible genetic disorders associated with a specific clinical phenotype. Targeted gene panels involve detailed analysis of a few genes, so they are less expensive than WES and easy to interpret; however, selecting the panel can be challenging (Quaio et al., 2021).

The prenatal sequencing comes with practical and ethical challenges impeding its implementation. First, as already seen in CMA, sequencing might uncover VOUS unrelated to the fetal/childhood health or nonsignificant. Secondly, technical and interpretation hurdles as a multidisciplinary team (geneticists, bio-informaticians, clinicians) are usually involved in interpreting and deciding what to report. Also, while WES may miss certain variants (e.g., deep intronic variants, structural variants, low-level mosaicism), WGS is complex for interpretation. Despite these challenges, as the sequencing will be cheaper and faster, it will be regarded as a first-tier diagnostic test, potentially bypassing the traditional stepwise approach of single-gene or panel testing. Global initiatives continue to accelerate the discovery of novel disease-causing genes through large-scale data sharing.

4. Clinical implications: Management and interventions

Identifying a genetic disorder in a fetus carries significant clinical implications for the management of the pregnancy, the delivery, and the interventions needed for both the baby and the mother. Here, we explore how genetic diagnoses influence pregnancy outcomes and decision-making for interventions to be taken.

4.1 Influence on parental decision-making

A prenatal genetic diagnosis often forces prospective parents to confront difficult choices. Many severe genetic disorders have prognoses that include perinatal death or profound disability. When such a condition is diagnosed, the families need to choose whether to continue the pregnancy with planned palliative care or opt for termination (TOP). The decision depends on both disease severity and detection time. One study showed a significant association between a fetal cardiac screening program and postnatal outcomes attributed to timely detection of severe forms of CHD and significant comorbidities, which also resulted in an increased TOP rate (Tomek et al., 2023). An early diagnosis allows families to align care as per their values and conditions. Like in Down syndrome (trisomy 21), prenatal diagnosis gives parents time to learn about the condition, connect with support groups, and prepare emotionally and logistically. This allows parents to decide and plan better without any surprises at birth. As seen recently, because of early diagnosis, some will choose to continue, others may not; in either case, having the diagnosis avoids surprises at birth and allows for better planning. Studies have shown a significant reduction in the birth prevalence of Down syndrome in some European countries, driven by high rates of TOP following a positive diagnosis (de Graaf et al., 2021).

4.2 Obstetric management adjustments

A known fetal genetic disorder often prompts adjustments in obstetric care with close monitoring for high-risk cases. For example, aneuploidies such as trisomy 18 and trisomy 13 are associated with fetal growth restriction (FGR) and placental dysfunction and will need monitoring of growth and amniotic fluid closely, as severe FGR could prompt early delivery. On the other hand, in Beckwith–Wiedemann syndrome, an imprinting disorder, there is a higher risk of fetal overgrowth and polyhydramnios, along with the chance of fetal hypoglycemia at birth due to hyperinsulinism; growth and glucose level monitoring are needed. Also, knowing the diagnosis leads to targeted surveillance and risk management along with planning timing and mode of delivery to optimize neonatal outcome. For example, if a fetus has spina bifida (myelomeningocele) or has a large omphalocele with Beckwith–Wiedemann, a planned caesarean at a speciality center might be needed to reduce complications and allow immediate surgical intervention. Similarly, for Turner syndrome (45, X) with an elevated risk of fetal demise in the third trimester, some clinicians consider delivery at 35–36 weeks, rather than continuing to 40 weeks.

4.3 Neonatal and pediatric outcomes

The genetic diagnosis shapes neonatal management postdelivery. Many genetic disorders come with known or predictable neonatal complications that can be managed if known before. Babies with DiGeorge syndrome (22q11.2 deletion) often suffer from hypocalcaemia in the neonatal period due to parathyroid hypoplasia, and after diagnosis, neonatologists will proactively monitor calcium and avoid neonatal seizures. Similarly, infants with congenital myotonic dystrophy (a trinucleotide repeat disorder) can have respiratory failure at birth and on diagnosis, the obstetric and neonatal teams can be ready to provide immediate respiratory support and plan for potential long-term ventilation needs. Metabolic disorders are well-suited to represent the advantages of early diagnosis, as many of them are treatable if recognised early. Like a baby with propionic acidemia or medium-chain acyl-CoA dehydrogenase (MCAD) deficiency, the medical team can preemptively institute measures (such as IV glucose, avoidance of certain feeds) right after birth to prevent the metabolic crisis that typically would otherwise land the baby in the NICU by day 3–5 of life. For treatable conditions such as CHD, a prenatal diagnosis ensures that the necessary specialists (surgeons, cardiologists, etc.) and support teams are on standby at delivery. Thus, comprehensive planning improves outcomes and reduces the chaos at birth.

5. Emerging therapeutic strategies

Compared with the past, where the focus was on diagnostics, recent years have seen remarkable progress in developing therapies targeting genetic disorders during fetal development. The emerging therapeutic strategies range from pharmacological treatments given to the pregnant mother (with the drug crossing into the fetus), to fetal surgeries and interventions, to cutting-edge approaches such as in utero gene therapy and genome editing.

5.1 In utero enzyme replacement and substrate reduction therapies

One class of treatable genetic diseases are the lysosomal storage disorders (LSDs) that are metabolic diseases and result from genetic enzyme deficiencies. Some LSDs, such as Pompe disease (glycogen storage type II), MPS VII (Sly syndrome), and MPS I (Hurler syndrome), have presented in utero with hydrops, indicating the pathological process starts before birth (Derderian et al., 2015). Enzyme replacement therapy (ERT) after birth has transformed outcomes for some of these disorders; for example, Al glucosidase alfa enzyme replacement therapy (ERT) using recombinant human GAA (rhGAA ERT) for infantile Pompe disease significantly improves survival and cardiac function (Nicolino et al., 2009). Researchers have reasoned that starting ERT even earlier, during pregnancy, that is, in utero enzyme replacement therapy (IUERT), might further improve outcomes because of a permissive blood-brain barrier. Preclinical studies in animal models of LSDs have shown encouraging results: administering enzymes to pregnant mice with fetuses that have LSDs can reduce storage material accumulation and improve neonatal condition (Nguyen et al., 2020).

Recently, IUERT was given to a fetus diagnosed with Pompe disease in a mother with fourth effected pregnancy. IUERT was started at 24 weeks of gestation, and the mother was given regular infusions of the enzyme (AL glucosidase alfa). The baby was born with much better outcomes, with no cardiac hypertrophy and minimal phenotypic and lifespan variation than the untreated sibling (Cohen et al., 2022). Similarly, there are ongoing clinical trials for in utero enzyme replacement in MPS I(NCT03580083)/MPS II(NCT03566043, NCT04571970, and NCT04597385). The rationale is to prevent the neurodegeneration and skeletal changes that even early postnatal bone marrow transplant and ERT cannot fully reverse if started after birth.

5.2 Fetal surgery and transfusion for genetic conditions

Fetal surgery is currently seen as a strategic approach for mitigating the progression of symptoms or correcting structural anomalies before they result in irreversible damage for certain anatomical problems. One of the examples includes myelomeningocele repair for spina bifida, which significantly improves outcomes by repairing the spine around 24 weeks. Spina bifida itself is usually not "genetic" but can be associated with some syndromes, and in utero surgical repair can help before major anatomical issues. Another condition where an in-utero procedure is seen to help is CDH, which can be seen isolated or in genetic syndromes (such as Cornelia de Lange or Fryns syndrome). Fetal endoscopic tracheal occlusion (FETO) has been developed to improve lung growth in severe CDH by placing a balloon in the fetus's trachea to block fluid egress, promoting lung growth (Deprest et al., 2004). Whether genetic or not, early trials show improved survival in severe CDH (Deprest et al., 2005).

Another emerging area is In-utero blood transfusion (IUT) and transplantation, such as for fetuses with genetic hematologic disorders, such as alpha thalassemia major (Hb Bart's hydrops fetalis), multiple in-utero blood transfusions have increased survival. The current approach is serial intrauterine transfusions to treat anemia and hydrops, which enable near-term delivery (around 36 weeks) while supporting normal neurodevelopment (Schwab et al., 2023). Recently, a phase 1 clinical trial for in utero hematopoietic stem cell transplantation (HSCT) for treating alpha thalassemia fetus have been started (NCT05797272). The idea is to attempt to cure the disease by transplanting stem cells from the mother (HLA match) into the fetus's peritoneal cavity after some transfusions. Researchers are pursuing in utero HSCT for other conditions, including immunodeficiencies (such as SCID) and metabolic diseases, with the hope that early transplant can prevent organ damage and also induce central tolerance, conceivably avoiding the need for immune suppression (since the fetal immune system can learn to accept donor cells as self) (Anchoo et al., 2025).

5.3 In utero gene therapy (IUGT)

Perhaps the most exciting and futuristic approach to treating congenital genetic disorders during fetal development is using IUGT. Fetal gene therapy includes delivery of functional genetic material through gene replacement vectors (such as AAV viral vectors) or the modification of gene expression through gene editing machinery (such as CRISPR) to the fetus to correct or mitigate genetic disorders before birth. There are certain advantages associated with IUGT: (1) The immune system is immature, potentially allowing viral vectors to persist without a limiting immune response and with the induction of antigen-specific immune tolerance (Chan et al., 2019; Colletti et al., 2008). (2) The small size of the fetus and target organs maximizes delivery and facilitates efficient gene transduction (Peranteau & Flake, 2020). (3) Early correction might prevent disease manifestations entirely. Various animal studies have shown the success of IUGT for diseases such as hemophilia, Gaucher's disease, and neurodevelopmental diseases (Chan et al., 2019; Massaro et al., 2020).

In gene additions, a functional copy of a gene through viral (e.g., AAV, lentivirus, adenovirus) or nonviral (e.g., lipid nanoparticles) vectors is introduced. Murine and nonhuman primate (NHP) models of hemophilia using AAV vectors have achieved long-term, therapeutic-level expression of clotting factors (FVIII, FIX), induced immune tolerance, and prevented bleeding phenotypes (Mattar et al., 2017; Sabatino et al., 2007). In mouse models of hemophilia (Factor VIII or IX deficiency), in utero injection of AAV vectors encoding the missing clotting factor resulted in pups being born with near-normal clotting and sustained expression of the factor and reducing bleeding episodes (Sabatino et al., 2007). Lentiviral vectors-IUGT delivering globin genes in a humanized mouse model of severe β-thalassemia major have shown phenotypic correction, including improved hematological indices and reduced extra-medullary hematopoiesis and iron deposition (Shangaris et al., 2019). Massaro et al. treated murine models of neuronopathic Gaucher disease (nGD) with IUGT by a single intracranial injection of human AAV9-GBA, resulting in normal phenotypic neuronal glucocerebrosidase expression. The treated mice with fetal IUGT recipients survived 18 weeks and showed no neurodegeneration and neuroinflammation compared with the untreated mice model, which exhibited fatal neurodegeneration and severely truncated lifespan.

Another disease amenable to IUGT is spinal muscular atrophy (SMA), which affects 8–10/100,000 newborns globally per annum, caused by mutation or deletion of the survival motor neuron 1 gene (*SMN1*) encoding SMN protein. SMN protein is needed for multiple essential cellular homeostatic and bioenergetic pathways (Chaytow et al., 2018). FDA-approved gene therapy includes AAV9-based SMN1 gene delivery given postnatally, sold commercially as Zolgensma, is seen to dramatically improve outcomes if given early in life; however, with high costs and relapses. The IUGT dose is believed to potentially help patients as a single injection of AAV9 intrauterine in the SMA mouse embryo improved life span and neuromuscular pathology (Rashnonejad et al., 2019).

5.4 Genome editing (CRISPR) in embryos/fetuses

Genome editing in human embryos using CRISPR-Cas9 holds immense potential for treating genetic diseases, especially monogenic diseases. However, the technique is both controversial and technically challenging. CRISPR/Cas9 triggers a nonhomologous end joining DNA repair pathway after DSB induction, resulting in insertion, deletion, substitutions, with loss-of-function, gain-of-function, and other unknown consequences. However, the conventional CRISPR/Cas9 is error-prone with off-targeting, mosaicism and chromosome deletion, raising major safety concerns. Ma et al. reported correction of the heterozygous MYBPC3 mutation in human preimplantation embryos using CRISPR-Cas9. MYBPC3 mutation causes hypertrophic cardiomyopathy (HCM) and affects 1 in 500 people (Ma et al., 2017). The study reported no evidence of mosaicism; however was later questioned for the detection methods used (Egli et al., 2018). Other studies have also revealed unwanted outcomes of CRISPR/Cas9 genome editing, such as deletions, rearrangements, and chromosomal abnormalities near the target locus and beyond in human embryos (Alanis-Lobato et al., 2021; Zuccaro et al., 2020). Although multiple improvements to CRISPR technology have been developed, including high-fidelity Cas9 variants that demonstrate reduced off-target activity (Kosicki et al., 2018).

Germline editing remains controversial and prohibited because of the complexity and possibility of unintended editing outcomes.

Recently, somatic gene editing has seen clinical success in β-thalassemia and sickle cell disease in a study done by Frangoul et al. (2021). The patients were injected with CRISPR/Cas9 edited $CD34^+$ hematopoietic stem and progenitor cells (HSPCs) after myeloablative conditioning. BCL11A is a transcription factor that represses γ-globin expression and fetal hemoglobin in erythroid cells and was edited in HSCPCs. Both patients showed a sustained increase in fetal hemoglobin distributed across red blood cells even after a year. The therapy is now approved and is marketed as *Casgevy* by CRISPR and *Vertex Pharma* in the UK and US (Wong, 2023).

Base editors (BE) and prime editing are newer classes of genome tools based on the CRISPR/Cas system that have been developed with high efficiency and no DSBs (Park et al., 2021; Song et al., 2020). These tools have been tried for gene therapy, construction of animal models and analyzing gene function (Lee et al., 2018; Zafra et al., 2018). The demand for BEs necessitates optimization and improvement, including a decrease in unwanted byproducts, enhancing editing activity and precision, and expanding the target range (Doman et al., 2020; Kim et al., 2017). mtDNA base editing approaches such as DddA-derived Cytosine Base Editors (DdCBEs) and mitochondrial-Targeted TALENs (mitoTALENs) are new and promising approaches to precisely treat mitochondrial diseases (Cho et al., 2022; Hashimoto et al., 2015). These approaches are protein-based and do not require RNA, as they cannot be efficiently delivered into mitochondria, making them ideal for the mitochondrial matrix. Currently, mitochondrial replacement therapy (MRT), that is, swapping out diseased mitochondria by transferring a nucleus into a donor egg, has already been applied in humans in a few cases. MRT is a form of germline intervention to prevent transmission of mtDNA disease and not a gene editing technique. The "three-parent baby" technique used involves fusion of nuclear DNA from parents and mtDNA from a matrix (Adashi et al., 2019). Ethically, MRT is considered acceptable under strict regulation in some countries.

6. Ethical, social, and policy considerations

As advances in prenatal genetic testing and therapies accelerate, there is an increase in complex ethical and social dilemmas as they influence intimate choices about starting a family, understanding children's lives with disabilities, and selecting appropriate medical interventions before birth. Thus, it is crucial to consider how and when we apply these tools to get maximum benefit and a commitment to the well-being of all.

6.1 Informed consent and counseling

As genomic data is complex, genetic counseling helps interpret results correctly and helps prospective parents to make informed choices about prenatal genetics. While offering NIPT to a low-risk population, it has to be conveyed to the patient that while NIPT has high accuracy for certain conditions, it is not a diagnostic technique, and an invasive confirmation may be needed in case of a positive (Rose & Kaimal, 2020). Also, the patients need to be informed about the conditions that can be screened or identified for every technique. The ACOG guidelines explicitly emphasize that testing should be an informed patient choice, considering the patient's values and needs (Rose & Kaimal, 2020). Also, all patients should be offered the testing along with counseling about the implications. The counseling should be balanced and nondirective while following guidelines or laws.

7. Selective reproduction and disability ethics

Prenatal detection of genetic conditions raises complex ethical questions about continuing pregnancy or abortion, which sometimes sends a negative message about disabled kids. Parents often make decisions due to the child's lifelong suffering and their capabilities to provide care, but this seems to make a worrying statement. As discussed before some European countries, the live birth of Down syndrome has fallen by 90% and is seen as "Eugenics," that is, the practice of selecting or eliminating traits within a population, often to improve genetic qualities. This leads to debate whether the diagnosis of certain conditions is ethical or not. However, the consensus is that healthcare professionals provide details along with nondirective counseling and support the decisions of the parents, whatsoever.

7.1 Equity in access to genomic medicine

A major social concern, access to technological advances such as NIPT, carrier screening, and diagnostic testing is not uniform and may increase already wide health disparities. NIPT remains free for many countries in Europe, while in other countries, it might be accessible only for those who can afford it. This raises an ethical issue of justice to access technology, as it gives rich people an added advantage to detect and an opportunity to avoid genetic conditions and associated complications. Currently, a publicly funded NIPT for all is advocated as it improves outcomes and reduces costs by managing undiagnosed conditions at birth. This also includes making genomic libraries and tests locally, as currently they are being validated in Europe or US cohorts. Countries such as the Gulf have unique allele frequencies, SNPs and mutations along with unique practices such as consanguinity. Yet it is underrepresented in the global database, which may lead to misrepresented and missed results. Although efforts are made to diversify the genomic reference dataset, tools are made to assess biased PRS and variant pathogenicity. This can be achieved by investment in local research data sharing and international collaborations.

7.2 Regulation of advanced technologies

CRISPR-edited babies in 2018 raised an alarm toward the use of gene editing in humans, resulting in the need for a global registry and regulation of gene editing experiments. As per consensus, germline editing should not be attempted clinically due to unknown risks. However, somatic gene editing (not heritable) is in trials for adults in some diseases. When it comes to fetuses, any form of gene editing is effectively germline if the germ cells (ovaries/testes) are edited too. Policies need to be developed based on societal and religious views as well as genetic makeup. Some communities may and are embracing these technologies to alleviate suffering, while others may fear they promote a "designer baby" mentality, which can cause unethical practices and reduce diversity. Even some countries don't allow any fetal modifications because of religious issues. Most bioethicists clearly distinguish between therapeutic (restoring health) and enhancement (improving beyond the norm) uses of gene editing, advocating regulation and policy development to prevent nonmedical or trivial applications.

References

Abozaid, G. M., Kerr, K., Alomary, H., Al-Omar, H. A., & McKnight, A. (2025). Global insight into rare disease and orphan drug definitions: a systematic literature review. *BMJ open, 15*(1), e086527. https://doi.org/10.1136/bmjopen-2024-086527

Adashi, E. Y., Caplan, A. L., Capron, A., Chapman, A. R., Cho, M., Clayton, E. W., Cohen, I. G., Cook-Deegan, R., Faden, R. R., Friedmann, T., Gostin, L. O., Greely, H. T., Johnston, J., Juengst, E., King, P. A., Knowles, L. P., Lyerly, A. D., McGuire, A. L., Moreno, J. D., … Walters, L. R. (2019). In support of mitochondrial replacement therapy. *Nature Medicine, 25*(6), 870–871. https://doi.org/10.1038/s41591-019-0477-4

Alanis-Lobato, G., Zohren, J., McCarthy, A., Fogarty, N. M. E., Kubikova, N., Hardman, E., Greco, M., Wells, D., Turner, J. M. A., & Niakan, K. K. (2021). Frequent loss of heterozygosity in CRISPR-Cas9-edited early human embryos. *Proceedings of the National Academy of Sciences of the United States of America, 118*(22). https://doi.org/10.1073/pnas.2004832117

Alfirevic, Z., Navaratnam, K., & Mujezinovic, F. (2017). Amniocentesis and chorionic villus sampling for prenatal diagnosis. *Cochrane Database of Systematic Reviews, 9*, Article CD003252.

Anchoo, C., Lev, A., Simon, A. J., Levy, S., Mandola, A., Frizinsky, S., Somekh, I., NaserEddin, A., Adwan, R. S. A., Toren, A., Golan, H., Bielorai, B., Hutt, D., Adam, E., & Somech, R. (2025). Outcome of hematopoietic stem cell transplantation for severe combined immunodeficiency and impact of newborn screening on overall survival: A single referral center study. *Journal of Allergy and Clinical Immunology, 155*(6), 1800–1812. https://doi.org/10.1016/j.jaci.2025.01.029

Beygo, J., Citro, V., Sparago, A., De crescenzo, A., Cerrato, F., Heitmann, M., Rademacher, K., Guala, A., Enklaar, T., Anichini, C., Cirillo silengo, M., Graf, N., Prawitt, D., Cubellis, M. V., Horsthemke, B., Buiting, K., & Riccio, A. (2013). The molecular function and clinical phenotype of partial deletions of the IGF2/H19 imprinting control region depends on the spatial arrangement of the remaining CTCF-binding sites. *Human Molecular Genetics, 22*(3), 544–557. https://doi.org/10.1093/hmg/dds465

Bianchi, D. W., Chudova, D., Sehnert, A. J., Bhatt, S., Murray, K., Prosen, T. L., Garber, J. E., Wilkins-Haug, L., Vora, N. L., Warsof, S., Goldberg, J., Ziainia, T., & Halks-Miller, M. (2015). Noninvasive prenatal testing and incidental detection of occult maternal malignancies. *JAMA, the Journal of the American Medical Association, 314*(2), 162–169. https://doi.org/10.1001/jama.2015.7120

Brooks, K. E., Daughtry, B. L., Davis, B., Yan, M. Y., Fei, S. S., Shepherd, S., Carbone, L., & Chavez, S. L. (2022). Molecular contribution to embryonic aneuploidy and karyotypic complexity in initial cleavage divisions ofmammalian development. *Development, 149*(7). https://doi.org/10.1242/dev.198341. Available from: https://journals.biologists.com/dev/article/149/7/dev198341/275037/Molecular-contribution-to-embryonic-aneuploidy-and

Cerrato, F., Sparago, A., Verde, G., De Crescenzo, A., Citro, V., Cubellis, M. V., Rinaldi, M. M., Boccuto, L., Neri, G., Magnani, C., D'Angelo, P., Collini, P., Perotti, D., Sebastio, G., Maher, E. R., & Riccio, A. (2008). Different mechanisms cause imprinting defects at the IGF2/H19 locus in Beckwith-Wiedemann syndrome and Wilms' tumour. *Human Molecular Genetics, 17*(10), 1427–1435. https://doi.org/10.1093/hmg/ddn031

Chan, J. K. Y., Gil-Farina, I., Johana, N., Rosales, C., Tan, Y. W., Ceiler, J., Mcintosh, J., Ogden, B., Waddington, S. N., Schmidt, M., Biswas, A., Choolani, M., Nathwani, A. C., & Mattar, C. N. Z. (2019). Therapeutic expression of human clotting factors IX and × following adeno-associated viral vector-mediated intrauterine gene transfer in early-gestation fetal macaques. *FASEB Journal, 33*(3), 3954–3967. https://doi.org/10.1096/fj.201801391R

Chandler, N. J., Ahlfors, H., Drury, S., Mellis, R., Hill, M., McKay, F. J., Collinson, C., Hayward, J., Jenkins, L., & Chitty, L. S. (2020). Noninvasive prenatal diagnosis for cystic fibrosis: Implementation, uptake, outcome, and implications. *Clinical Chemistry, 66*(1), 207–216. https://doi.org/10.1373/clinchem.2019.305011

Chavez, S. L., Loewke, K. E., Han, J., Moussavi, F., Colls, P., Munne, S., Behr, B., & Reijo Pera, R. A. (2012). Dynamic blastomere behaviour reflects human embryo ploidy by the four-cell stage. *Nature Communications, 3*. https://doi.org/10.1038/ncomms2249. Available from: https://www.nature.com/ncomms/

Chaytow, H., Huang, Y. T., Gillingwater, T. H., & Faller, K. M. E. (2018). The role of survival motor neuron protein (SMN) in protein homeostasis. *Cellular and Molecular Life Sciences, 75*(21), 3877–3894. https://doi.org/10.1007/s00018-018-2849-1

Chekroun, I., Shenbagam, S., Almarri, M. A., Mokrab, Y., Uddin, M., Alkhnbashi, O. S., Zaki, M. S., Najmabadi, H., Kahrizi, K., Fakhro, K. A., Almontashiri, N. A. M., Ali, F. R., Özbek, U., Reversade, B., Alkuraya, F. S., Alsheikh-Ali, A., & Abou Tayoun, A. N. (2025). Genomics of rare diseases in the Greater Middle East. *Emirates Nature Genetics, 57*(3), 505–514. https://doi.org/10.1038/s41588-025-02075-8

Chitty, L. S., Mason, S., Barrett, A. N., Mckay, F., Lench, N., Daley, R., & Jenkins, L. A. (2015). Non-invasive prenatal diagnosis of achondroplasia and thanatophoric dysplasia: Next-generation sequencing allows for a safer, more accurate, and comprehensive approach. *Prenatal Diagnosis, 35*(7), 656–662. https://doi.org/10.1002/pd.4583. Available from: http://onlinelibrary.wiley.com/journal/10.1002/(ISSN)1097-0223

Chiu, R. W. K., Akolekar, R., Zheng, Y. W. L., Leung, T. Y., Sun, H., Chan, K. C. A., Lun, F. M. F., Go, A. T. J. I., Lau, E. T., To, W. W. K., Leung, W. C., Tang, R. Y. K., Au-Yeung, S. K. C., Lam, H., Kung, Y. Y., Zhang, X., van Vugt, J. M. G., Minekawa, R., Tang, M. H. Y., … Lo, Y. M. D. (2011). Non-invasive prenatal assessment of trisomy 21 by multiplexed maternal plasma DNA sequencing: Large scale validity study. *BMJ, 342*(jan11 1), Article c7401. https://doi.org/10.1136/bmj.c7401

Cho, S. I., Lee, S., Mok, Y. G., Lim, K., Lee, J., Lee, J. M., Chung, E., & Kim, J. S. (2022). Targeted A-to-G base editing in human mitochondrial DNA with programmable deaminases. *Cell, 185*(10), 1764. https://doi.org/10.1016/j.cell.2022.03.039

Christiaens, L., Chitty, L. S., & Langlois, S. (2021). 'Current controversies in prenatal diagnosis: Expanded NIPT that includes conditions other than trisomies 13, 18, and 21 should be offered. *Prenatal Diagnosis, 41*, 1316–1323.

Cohen, J. L., Chakraborty, P., Fung-Kee-Fung, K., Schwab, M. E., Bali, D., Young, S. P., Gelb, M. H., Khaledi, H., Dibattista, A., Smallshaw, S., Moretti, F., Wong, D., Lacroix, C., El Demellawy, D., Strickland, K. C., Lougheed, J., Moon-Grady, A., Lianoglou, B. R., Harmatz, P., Kishnani, P. S., & MacKenzie, T. C. (2022). In utero enzyme-replacement therapy for infantile-onset Pompe's disease. *New England Journal of Medicine, 387*(23), 2150–2158. https://doi.org/10.1056/NEJMoa2200587

Colletti, E., Lindstedt, S., Park, P. J., Almeida-Porada, G., & Porada, C. D. (2008). Early fetal gene delivery utilizes both central and peripheral mechanisms of tolerance induction. *Experimental Hematology, 36*(7), 816–822. https://doi.org/10.1016/j.exphem.2008.02.007

Condo, I. (2022). Rare monogenic diseases: Molecular pathophysiology and novel therapies. *International Journal of Molecular Sciences, 23*.

Cortessis, V. K., Azadian, M., Buxbaum, J., Sanogo, F., Song, A. Y., Sriprasert, I., Wei, P. C., Yu, J., Chung, K., & Siegmund, K. D. (2018). Comprehensive meta-analysis reveals association between multiple imprinting disorders and conception by assisted reproductive technology. *Journal of Assisted Reproduction and Genetics, 35*(6), 943–952. https://doi.org/10.1007/s10815-018-1173-x

D'Aversa, E., Breveglieri, G., Pellegatti, P., Guerra, G., Gambari, R., & Borgatti, M. (2018). Non-invasive fetal sex diagnosis in plasma of early weeks pregnants using droplet digital PCR. *Molecular Medicine, 24*, 14.

Dawood, M., Heavner, B., Wheeler, M. M., Ungar, R. A., LoTempio, J., Wiel, L., Berger, S., Bernstein, J. A., Chong, J. X., Délot, E. C., Eichler, E. E., Gibbs, R. A., Lupski, J. R., Shojaie, A., Talkowski, M. E., Wagner, A. H., Wei, C. L., Wellington, C., Wheeler, M. T., … Montgomery, S. B. (2024). GREGoR: Accelerating genomics for rare diseases. *arXiv*. https://doi.org/10.48550/arXiv.2412.14338. Available from https://arxiv.org

De Coster, T., Masset, H., Tšuiko, O., Catteeuw, M., Zhao, Y., Dierckxsens, N., Aparicio, A. L., Dimitriadou, E., Debrock, S., Peeraer, K., de Ruijter-Villani, M., Smits, K., Van Soom, A., & Vermeesch, J. R. (2022). Parental genomes segregate into distinct blastomeres during multipolar zygotic divisions leading to mixoploid and chimeric blastocysts. *Genome Biology, 23*(1). https://doi.org/10.1186/s13059-022-02763-2

de Graaf, G., Buckley, F., & Skotko, B. G. (2021). Estimation of the number of people with Down syndrome in Europe. *European Journal of Human Genetics, 29*, 402–410.

Deprest, J., Gratacos, E., Nicolaides, K. H., & Task Group, F. (2004). Fetoscopic tracheal occlusion (FETO) for severe congenital diaphragmatic hernia: Evolution of a technique and preliminary results. *Ultrasound in Obstetrics and Gynecology, 24*, 121–126.

Deprest, J., Jani, J., Gratacos, E., Vandecruys, H., Naulaers, G., Delgado, J., Greenough, A., Nicolaides, K., Van Schoubroeck, D., Vandevelde, M., Devlieger, H., Cannie, M., Dymarkowski, S., Lerut, T., Carreras, E., Salcedo, S., Toran, N., Peiro, J. L., Martinez-Ibanez, V., … Davenport, M. (2005). Fetal intervention for congenital diaphragmatic hernia: The European experience. *Seminars in Perinatology, 29*(2), 94–103. https://doi.org/10.1053/j.semperi.2005.04.006

Derderian, S. C., Jeanty, C., Fleck, S. R., Cheng, L. S., Peyvandi, S., Moon-Grady, A. J., Farrell, J., Hirose, S., Gonzalez, J., Keller, R. L., & MacKenzie, T. C. (2015). The many faces of hydrops. *Journal of Pediatric Surgery, 50*(1), 50–54. https://doi.org/10.1016/j.jpedsurg.2014.10.027

Destouni, A., Esteki, M. Z., Catteeuw, M., Tšuiko, O., Dimitriadou, E., Smits, K., Kurg, A., Salumets, A., Van Soom, A., Voet, T., & Vermeesch, J. R. (2016). Zygotes segregate entire parental genomes in distinct blastomere lineages causing cleavage-stage chimerism and mixoploidy. *Genome Research, 26*(5), 567–578. https://doi.org/10.1101/gr.200527.115

Devaney, S. A., Palomaki, G. E., Scott, J. A., & Bianchi, D. W. (2011). Noninvasive fetal sex determination using cell-free fetal DNA: A systematic review and meta-analysis. *JAMA, 306*(6), 627–636. https://doi.org/10.1001/jama.2011.1114

Dolanc Merc, M., Peterlin, B., & Lovrecic, L. (2023). The genetic approach to stillbirth: A systematic review. *Prenatal Diagnosis, 43*, 1220-28.

Doman, J. L., Raguram, A., Newby, G. A., & Liu, D. R. (2020). Evaluation and minimization of Cas9-independent off-target DNA editing by cytosine base editors. *Nature Biotechnology, 38*(5), 620–628. https://doi.org/10.1038/s41587-020-0414-6

Egli, D., Zuccaro, M. V., Kosicki, M., Church, G. M., Bradley, A., & Jasin, M. (2018). Inter-homologue repair in fertilized human eggs? *Nature, 560*(7717), Article E5. https://doi.org/10.1038/s41586-018-0379-5

El Naofal, M., Ramaswamy, S., Alsarhan, A., Nugud, A., Sarfraz, F., Janbaz, H., Taylor, A., Jain, R., Halabi, N., Yaslam, S., Alfalasi, R., Shenbagam, S., Rabea, F., Bitzan, M., Yavuz, L., Wafadari, D., Abulhoul, H., Shankar, S., Al Maazmi, M., … Abou Tayoun, A. N. (2023). The genomic landscape of rare disorders in the Middle East. *Emirates Genome Medicine, 15*(1). https://doi.org/10.1186/s13073-023-01157-8

Frangoul, H., Altshuler, D., Cappellini, M. D., Chen, Y. S., Domm, J., Eustace, B. K., Foell, J., De La Fuente, J., Grupp, S., Handgretinger, R., Ho, T. W., Kattamis, A., Kernytsky, A., Lekstrom-Himes, J., Li, A. M., Locatelli, F., Mapara, M. Y., De Montalembert, M., Rondelli, D., … Corbacioglu, S. (2021). CRISPR-Cas9 gene editing for sickle cell disease and β-thalassemia. *New England Journal of Medicine, 384*(3), 252–260. https://doi.org/10.1056/NEJMoa2031054

Gil, M. M., Accurti, V., Santacruz, B., Plana, M. N., & Nicolaides, K. H. (2017). Analysis of cell-free DNA in maternal blood in screening for aneuploidies: Updated meta-analysis. *Ultrasound in Obstetrics and Gynecology, 50*(3), 302–314. https://doi.org/10.1002/uog.17484

Gorman, G. S., Chinnery, P. F., DiMauro, S., Hirano, M., Koga, Y., McFarland, R., Suomalainen, A., Thorburn, D. R., Zeviani, M., & Turnbull, D. M. (2016). Mitochondrial diseases. *Nature Reviews Disease Primers, 2*. https://doi.org/10.1038/nrdp.2016.80. Available from: http://www.nature.com/nrdp/

Gregg, A. R., Skotko, B. G., Benkendorf, J. L., Monaghan, K. G., Bajaj, K., Best, R. G., Klugman, S., & Watson, M. S. (2016). Noninvasive prenatal screening for fetal aneuploidy, 2016 update: A position statement of the American College of Medical Genetics and Genomics. *Genetics in Medicine, 18*(10), 1056–1065. https://doi.org/10.1038/gim.2016.97

Hashimoto, M., Bacman, S. R., Peralta, S., Falk, M. J., Chomyn, A., Chan, D. C., Williams, S. L., & Moraes, C. T. (2015). MitoTALEN: A general approach to reduce mutant mtDNA loads and restore oxidative phosphorylation function in mitochondrial diseases. *Molecular Therapy, 23*(10), 1592–1599. https://doi.org/10.1038/mt.2015.126

Hayashi, S., & Umeda, T. (2008). 35 years of Japanese policy on rare diseases. *The Lancet, 372*(9642), 889–890. https://doi.org/10.1016/S0140-6736(08)61393-8

Hyde, K. J., & Schust, D. J. (2015). 'Genetic considerations in recurrent pregnancy loss. *Cold Spring Harbor Perspectives in Medicine, 5*, Article a023119.

Jaslow, C. R., Carney, J. L., & Kutteh, W. H. (2010). 'Diagnostic factors identified in 1020 women with two versus three or more recurrent pregnancy losses. *Fertility and Sterility, 93*, 1234–1243.

Jayashankar, S. S., Nasaruddin, M. L., Hassan, M. F., Dasrilsyah, R. A., Shafiee, M. N., Ismail, N. A. S., & Alias, E. (2023). Non-invasive prenatal testing (NIPT): Reliability, challenges, and future directions. *Diagnostics, 13*(15). https://doi.org/10.3390/diagnostics13152570. Available from: http://www.mdpi.com/journal/diagnostics/

Kim, Y. B., Komor, A. C., Levy, J. M., Packer, M. S., Zhao, K. T., & Liu, D. R. (2017). Increasing the genome-targeting scope and precision of base editing with engineered Cas9-cytidine deaminase fusions. *Nature Biotechnology, 35*(4), 371–376. https://doi.org/10.1038/nbt.3803

Kosicki, M., Tomberg, K., & Bradley, A. (2018). Repair of double-strand breaks induced by CRISPR-Cas9 leads to large deletions and complex rearrangements. *Nature Biotechnology, 36*, 765–771.

Kuhlmann-Capek, M., Chiossi, G., Singh, P., Monsivais, L., Lozovyy, V., Gallagher, L., Kirsch, N., Florence, E., Petruzzi, V., Chang, J., Buenaventura, S., Walden, P., Gardner, B., Munn, M., & Costantine, M. (2019). Effects of medication intake in early pregnancy on the fetal fraction of cell-free DNA testing. *Prenatal Diagnosis, 39*(5), 361–368. https://doi.org/10.1002/pd.5436. Available from: http://obgyn.onlinelibrary.wiley.com/hub/journal/10.1002/(ISSN)1097-0223/

Lee, C. E., Singleton, K. S., Wallin, M., & Faundez, V. (2020). Rare genetic diseases: Nature's experiments on human development. *iScience, 23*(5). https://doi.org/10.1016/j.isci.2020.101123. Available from: www.cell.com/iscience

Lee, H. K., Willi, M., Miller, S. M., Kim, S., Liu, C., Liu, D. R., & Hennighausen, L. (2018). Targeting fidelity of adenine and cytosine base editors in mouse embryos. *Nature Communications, 9*(1). https://doi.org/10.1038/s41467-018-07322-7. Available from https://www.nature.com/ncomms/

Legler, T. J., Lührig, S., Korschineck, I., & Schwartz, D. (2021). Diagnostic performance of the noninvasive prenatal FetoGnost RhD assay for the prediction of the fetal RhD blood group status. *Archives of Gynecology and Obstetrics, 304*(5), 1191–1196. https://doi.org/10.1007/s00404-021-06055-1

Levy, B., & Wapner, R. (2018). Prenatal diagnosis by chromosomal microarray analysis. *Fertility and Sterility, 109*(2), 201–212. https://doi.org/10.1016/j.fertnstert.2018.01.005

Lo, Y. M., Corbetta, N., Chamberlain, P. F., Rai, V., Sargent, I. L., Redman, C. W., & Wainscoat, J. S. (1997). Presence of fetal DNA in maternal plasma and serum. *Lancet, 350*, 485–487.

Lo, Y. M. D., Lun, F. M. F., Chan, K. C. A., Tsui, N. B. Y., Chong, K. C., Lau, T. K., Leung, T. Y., Zee, B. C. Y., Cantor, C. R., & Chiu, R. W. K. (2007). Digital PCR for the molecular detection of fetal chromosomal aneuploidy. *Proceedings of the National Academy of Sciences of the United States of America, 104*(32), 13116–13121. https://doi.org/10.1073/pnas.0705765104

Lord, J., McMullan, D. J., Eberhardt, R. Y., Rinck, G., Hamilton, S. J., Quinlan-Jones, E., Prigmore, E., Keelagher, R., Best, S. K., Carey, G. K., Mellis, R., Robart, S., Berry, I. R., Chandler, K. E., Cilliers, D., Cresswell, L., Edwards, S. L., Gardiner, C., Henderson, A., … Wilson, E. (2019). Prenatal exome sequencing analysis in fetal structural anomalies detected by ultrasonography (PAGE): A cohort study. *The Lancet, 393* (10173), 747–757. https://doi.org/10.1016/S0140-6736(18)31940-8

Ma, H., Marti-Gutierrez, N., Park, S. W., Wu, J., Lee, Y., Suzuki, K., Koski, A., Ji, D., Hayama, T., Ahmed, R., Darby, H., Van Dyken, C., Li, Y., Kang, E., Park, A. R., Kim, D., Kim, S. T., Gong, J., Gu, Y., … Mitalipov, S. (2017). Correction of a pathogenic gene mutation in human embryos. *Nature, 548*(7668), 413–419. https://doi.org/10.1038/nature23305

Massaro, G., Hughes, M. P., Whaler, S. M., Wallom, K. L., Priestman, D. A., Platt, F. M., Waddington, S. N., & Rahim, A. A. (2020). Systemic AAV9 gene therapy using the synapsin I promoter rescues a mouse model of neuronopathic Gaucher disease but with limited cross-correction potential to astrocytes. *Human Molecular Genetics, 29*(12), 1933–1949. https://doi.org/10.1093/HMG/DDZ317

Masset, H., Tsuiko, O., & Vermeesch, J. R. (2021). 'Genome-wide abnormalities in embryos: Origins and clinical consequences. *Prenatal Diagnosis, 41*, 554–563.

Mathews, T. J., & Driscoll, A. K. (2017). Trends in infant mortality in the United States, 2005–2014. *NCHS data brief, 279*, 1–8.

Mattar, C. N. Z., Gil-Farina, I., Rosales, C., Johana, N., Tan, Y. Y. W., McIntosh, J., Kaeppel, C., Waddington, S. N., Biswas, A., Choolani, M., Schmidt, M., Nathwani, A. C., & Chan, J. K. Y. (2017). In utero transfer of adeno-associated viral vectors produces long-term factor IX levels in a cynomolgus macaque model. *Molecular Therapy, 25*(8), 1843–1853. https://doi.org/10.1016/j.ymthe.2017.04.003

Meng, L., Pammi, M., Saronwala, A., Magoulas, P., Ghazi, A. R., Vetrini, F., Zhang, J., He, W., Dharmadhikari, A. V., Qu, C., Ward, P., Braxton, A., Narayanan, S., Ge, X., Tokita, M. J., Santiago-Sim, T., Dai, H., Chiang, T., Smith, H., … Lalani, S. R. (2017). Use of exome sequencing for infants in intensive care units ascertainment of severe single-gene disorders and effect on medical management. *JAMA Pediatrics, 171*(12). https://doi.org/10.1001/jamapediatrics.2017.3438. Available from: https://jamanetwork.com/journals/jamapediatrics/articlepdf/2654887/jamapediatrics_meng_2017_oi_170076.pdf

Miller, D. T., Adam, M. P., Aradhya, S., Biesecker, L. G., Brothman, A. R., Carter, N. P., Church, D. M., Crolla, J. A., Eichler, E. E., Epstein, C. J., Faucett, W. A., Feuk, L., Friedman, J. M., Hamosh, A., Jackson, L., Kaminsky, E. B., Kok, K., Krantz, I. D., Kuhn, R. M., … Ledbetter, D. H. (2010). Consensus statement: Chromosomal microarray is a first-tier clinical diagnostic test for individuals with developmental disabilities or congenital anomalies. *The American Journal of Human Genetics, 86*(5), 749–764. https://doi.org/10.1016/j.ajhg.2010.04.006

Minear, M. A., Lewis, C., Pradhan, S., & Chandrasekharan, S. (2015). Global perspectives on clinical adoption of NIPT. *Prenatal Diagnosis, 35*(10), 959–967. https://doi.org/10.1002/pd.4637. Available from: http://obgyn.onlinelibrary.wiley.com/hub/journal/10.1002/(ISSN)1097-0223/

Monk, D., Mackay, D. J. G., Eggermann, T., Maher, E. R., & Riccio, A. (2019). Genomic imprinting disorders: Lessons on how genome, epigenome and environment interact. *Nature Reviews Genetics, 20*(4), 235–248. https://doi.org/10.1038/s41576-018-0092-0

Munné, S. (2006). Chromosome abnormalities and their relationship to morphology and development of human embryos. *Reproductive BioMedicine Online, 12*(2), 234–253. https://doi.org/10.1016/s1472-6483(10)60866-8

Munne, S., Magli, C., Bahce, M., Fung, J., Legator, M., Morrison, L., Cohert, J., & Gianaroli, L. (1998). 'Preimplantation diagnosis of the aneuploidies most commonly found in spontaneous abortions and live births: XY, 13, 14, 15, 16, 18, 21, 22. *Prenatal Diagnosis, 18*, 1459–1466.

Nagamatsu, T., Kamei, Y., Yamashita, T., Fujii, T., & Kozuma, S. (2014). Placental abnormalities detected by ultrasonography in a case of confined placental mosaicism for trisomy 2 with severe fetal growth restriction. *Journal of Obstetrics and Gynaecology Research, 40* (1), 279–283. https://doi.org/10.1111/jog.12145

Nguyen, Q. H., Witt, R. G., Wang, B., Eikani, C., Shea, J., Smith, L. K., Boyle, G., Cadaoas, J., Sper, R., MacKenzie, J. D., Villeda, S., & MacKenzie, T. C. (2020). Tolerance induction and microglial engraftment after fetal therapy without conditioning in mice with mucopolysaccharidosis type VII. *Science Translational Medicine, 12*(532). https://doi.org/10.1126/scitranslmed.aay8980. Available from: https://stm.sciencemag.org/content/12/532/eaay8980

Nicolaides, K. H., Azar, G., Byrne, D., Mansur, C., & Marks, K. (1992). Fetal nuchal translucency: Ultrasound screening for chromosomal defects in first trimester of pregnancy. *BMJ, 304*(6831), 867–869. https://doi.org/10.1136/bmj.304.6831.867

Nicolino, M., Byrne, B., Wraith, J. E., Leslie, N., Mandel, H., Freyer, D. R., Arnold, G. L., Pivnick, E. K., Ottinger, C. J., Robinson, P. H., Loo, J. C. A., Smitka, M., Jardine, P., Tatò, L., Chabrol, B., McCandless, S., Kimura, S., Mehta, L., Bali, D., … Kishnani, P. S. (2009). Clinical outcomes after long-term treatment with alglucosidase alfa in infants and children with advanced Pompe disease. *Genetics in Medicine, 11*(3), 210–219. https://doi.org/10.1097/GIM.0b013e31819d0996

Norton, M. E., & Wapner, R. J. (2015). Cell-free DNA analysis for noninvasive examination of trisomy. *New England Journal of Medicine, 373*(26), 2582. https://doi.org/10.1056/NEJMc1509344

Palomaki, G. E., Kloza, E. M., Lambert-Messerlian, G. M., Haddow, J. E., Neveux, L. M., Ehrich, M., Van Den Boom, D., Bombard, A. T., Deciu, C., Grody, W. W., Nelson, S. F., & Canick, J. A. (2011). DNA sequencing of maternal plasma to detect Down syndrome: An international clinical validation study. *Genetics in Medicine, 13* (11), 913–920. https://doi.org/10.1097/GIM.0b013e3182368a0e

Park, S. J., Jeong, T. Y., Shin, S. K., Yoon, D. E., Lim, S. Y., Kim, S. P., Choi, J., Lee, H., Hong, J. I., Ahn, J., Seong, J. K., & Kim, K. (2021). Targeted mutagenesis in mouse cells and embryos using an enhanced prime editor. *Genome Biology, 22*(1). https://doi.org/10.1186/s13059-021-02389-w. Available from: http://genomebiology.com/

Parks, M., Court, S., Bowns, B., Cleary, S., Clokie, S., Hewitt, J., Williams, D., Cole, T., Macdonald, F., Griffiths, M., & Allen, S. (2017). Non-invasive prenatal diagnosis of spinal muscular atrophy by relative haplotype dosage. *European Journal of Human Genetics, 25* (4), 416–422. https://doi.org/10.1038/ejhg.2016.195

Parks, M., Court, S., Cleary, S., Clokie, S., Hewitt, J., Williams, D., Cole, T., Macdonald, F., Griffiths, M., & Allen, S. (2016). Non-invasive prenatal diagnosis of Duchenne and Becker muscular dystrophies by relative haplotype dosage. *Prenatal Diagnosis, 36*(4),

312–320. https://doi.org/10.1002/pd.4781. Available from: http://onlinelibrary.wiley.com/journal/10.1002/(ISSN)1097-0223

Peranteau, W. H., & Flake, A. W. (2020). The future of in utero gene therapy. *Molecular Diagnosis and Therapy, 24*(2), 135–142. https://doi.org/10.1007/s40291-020-00445-y

Petrovski, S., Aggarwal, V., Giordano, J. L., Stosic, M., Wou, K., Bier, L., Spiegel, E., Brennan, K., Stong, N., Jobanputra, V., Ren, Z., Zhu, X., Mebane, C., Nahum, O., Wang, Q., Kamalakaran, S., Malone, C., Anyane-Yeboa, K., Miller, R., … Wapner, R. J. (2019). Whole-exome sequencing in the evaluation of fetal structural anomalies: A prospective cohort study. *The Lancet, 393*(10173), 758–767. https://doi.org/10.1016/S0140-6736(18)32042-7

Platt, F. M., d'Azzo, A., Davidson, B. L., Neufeld, E. F., & Tifft, C. J. (2018). Lysosomal storage diseases. *Nature Reviews Disease Primers, 4*(1). https://doi.org/10.1038/s41572-018-0025-4. Available from: http://www.nature.com/nrdp/

Quaio, C. R. D. C., Obando, M. J. R., Perazzio, S. F., Dutra, A. P., Chung, C. H., Moreira, C. M., Filho, G. M. N., Sacramento-Bobotis, P. R., Penna, M. G., de Souza, R. R. F., Cintra, V. P., Carnavalli, J. E. P., da Silva, R. A., Santos, M. N. P., Paixão, D., Baratela, W. A.d. R., Olivati, C., Spolador, G. M., Pintao, M. C., … Kim, C. A. (2021). Exome sequencing and targeted gene panels: A simulated comparison of diagnostic yield using data from 158 patients with rare diseases. *Brazilian Journal of Genetics, Brazil Genetics and Molecular Biology, 44*(4). https://doi.org/10.1590/1678-4685-GMB-2021-0061. Available from: https://www.scielo.br/j/gmb/a/6CNmRWLwvGY3W7w37pj6nQw/?format=pdf&lang=en

Rashnonejad, A., Amini Chermahini, G., Gündüz, C., Onay, H., Aykut, A., Durmaz, B., Baka, M., Su, Q., Gao, G., & Özkınay, F. (2019). Fetal gene therapy using a single injection of recombinant AAV9 rescued SMA phenotype in mice. *Molecular Therapy, 27*(12), 2123–2133. https://doi.org/10.1016/j.ymthe.2019.08.017

Ravitsky, V., Roy, M. C., Haidar, H., Henneman, L., Marshall, J., Newson, A. J., Ngan, O. M. Y., & Nov-Klaiman, T. (2021). The emergence and global spread of noninvasive prenatal testing. *Annual Review of Genomics and Human Genetics, 22*, 309–338. https://doi.org/10.1146/annurev-genom-083118-015053

Reiss, R. E., Discenza, M., Foster, J., Dobson, L., & Wilkins-Haug, L. (2017). Sex chromosome aneuploidy detection by noninvasive prenatal testing: Helpful or hazardous? *Prenatal Diagnosis, 37*(5), 515–520. https://doi.org/10.1002/pd.5039. Available from: http://onlinelibrary.wiley.com/journal/10.1002/(ISSN)1097-0223

Rose, N. C., & Kaimal, A. J. (2020). Screening for fetal chromosomal abnormalities: ACOG practice bulletin summary, number 226. *Obstetrics & Gynecology, 136*, 859–861.

Rotig, A., de Lonlay, P. D., Chretien, D., Foury, F., Koenig, M., Sidi, D., Munnich, A., & Rustin, P. (1997). Aconitase and mitochondrial iron-sulphur protein deficiency in friedreich ataxia. *Nature Genetics, 17*(2), 215–217. https://doi.org/10.1038/ng1097-215

Sabatino, D. E., MacKenzie, T. C., Peranteau, W., Edmonson, S., Campagnoli, C., Liu, Y. L., Flake, A. W., & High, K. A. (2007). Persistent expression of hF.IX after tolerance induction by in utero or neonatal administration of AAV-1-F.IX in hemophilia B mice. *Molecular Therapy, 15*(9), 1677–1685. https://doi.org/10.1038/sj.mt.6300219

Saunders, C. J., Miller, N. A., Soden, S. E., Dinwiddie, D. L., Noll, A., Alnadi, N. A., Andraws, N., Patterson, M. L. A., Krivohlavek, L. A., Fellis, J., Humphray, S., Saffrey, P., Kingsbury, Z., Weir, J. C., Betley, J., Grocock, R. J., Margulies, E. H., Farrow, E. G., Artman, M., … Kingsmore, S. F. (2012). Rapid whole-genome sequencing for genetic disease diagnosis in neonatal intensive care units. *Science Translational*

Medicine, 4(154). https://doi.org/10.1126/scitranslmed.3004041. Available from: http://stm.sciencemag.org/content/4/154/154ra135.full.pdf

Schwab, M. E., Lianoglou, B. R., Gano, D., Velez, J. G., Allen, I. E., Arvon, R., Baschat, A., Bianchi, D. W., Bitanga, M., Bourguignon, A., Brown, R. N., Chen, B., Chien, M., Davis-Nelson, S., de Laat, M. W. M., Ekwattanakit, S., Gollin, Y., Hirata, G., Jelin, A., … MacKenzie, T. C. (2023). The impact of in utero transfusions on perinatal outcomes in patients with alpha thalassemia major: The UCSF registry. *Blood Advances, 7*(2), 269–279. https://doi.org/10.1182/bloodadvances.2022007823

Shangaris, P., Loukogeorgakis, S. P., Subramaniam, S., Flouri, C., Jackson, L. H., Wang, W., Blundell, M. P., Liu, S., Eaton, S., Bakhamis, N., Ramachandra, D. L., Maghsoudlou, P., Urbani, L., Waddington, S. N., Eddaoudi, A., Archer, J., Antoniou, M. N., Stuckey, D. J., Schmidt, M., … David, A. L. (2019). Utero gene therapy (IUGT) using GLOBE lentiviral vector phenotypically corrects the heterozygous humanised mouse model and its progress can be monitored using MRI techniques. *Scientific Reports, 9*(1). https://doi.org/10.1038/s41598-019-48078-4. Available from: www.nature.com/srep/index.html

Shaw, J., Scotchman, E., Chandler, N., & Chitty, L. S. (2020). Preimplantation genetic testing: Non-invasive prenatal testing for aneuploidy, copy-number variants and single-gene disorders. *Reproduction, 160*(5), A1. https://doi.org/10.1530/rep-19-0591

Singla, S., Iwamoto-Stohl, L. K., Zhu, M., & Zernicka-Goetz, M. (2020). Autophagy-mediated apoptosis eliminates aneuploid cells in a mouse model of chromosome mosaicism. *Nature Communications, 11*(1). https://doi.org/10.1038/s41467-020-16796-3. Available from: https://www.nature.com/ncomms/

Song, Y., Liu, Z., Zhang, Y., Chen, M., Sui, T., Lai, L., & Li, Z. (2020). Large-fragment deletions induced by Cas9 cleavage while not in the BEs system. *Molecular Therapy Nucleic Acids, 21*, 523–526. https://doi.org/10.1016/j.omtn.2020.06.019

Springett, A., Wellesley, D., Greenlees, R., Loane, M., Addor, M. C., Arriola, L., Bergman, J., Cavero-Carbonell, C., Csaky-Szunyogh, M., Draper, E. S., Garne, E., Gatt, M., Haeusler, M., Khoshnood, B., Klungsoyr, K., Lynch, C., Dias, C. M., Mcdonnell, R., Nelen, V., … Morris, J. K. (2015). Congenital anomalies associated with trisomy 18 or trisomy 13: A registry-based study in 16 European countries, 2000-2011. *American Journal of Medical Genetics, Part A, 167*(12), 3062–3069. https://doi.org/10.1002/ajmg.a.37355. Available from: http://onlinelibrary.wiley.com/journal/10.1002/(ISSN)1552-4833

Stewart, J. B., & Chinnery, P. F. (2015). The dynamics of mitochondrial DNA heteroplasmy: Implications for human health and disease. *Nature Reviews Genetics, 16*(9), 530–542. https://doi.org/10.1038/nrg3966

Tomek, V., Jicinska, H., Pavlicek, J., Kovanda, J., Jehlicka, P., Klaskova, E., Mrazek, J., Cutka, D., Smetanova, D., Brestak, M., Vlasin, P., Pavlikova, M., Chaloupecky, V., Janousek, J., & Marek, J. (2023). Pregnancy termination and postnatal major congenital heart defect prevalence after introduction of prenatal cardiac screening. *JAMA Network Open, 6*, Article e2334069.

Turki, R. F., Assidi, M., Banni, H. A., Zahed, H. A., Karim, S., Schulten, H. J., Abu-Elmagd, M., Rouzi, A. A., Bajouh, O., Jamal, H. S., Al-Qahtani, M. H., & Abuzenadah, A. M. (2016). Associations of recurrent miscarriages with chromosomal abnormalities, thrombophilia allelic polymorphisms and/or consanguinity in Saudi Arabia. *BMC Medical Genetics, 17*. https://doi.org/10.1186/s12881-016-0331-1. Available from http://www.biomedcentral.com/bmcmedgenet/

van den Berg, M. M. J., van Maarle, M. C., van Wely, M., & Goddijn, M. (2012). Genetics of early miscarriage. *Biochimica et Biophysica Acta—Molecular Basis of Disease, 1822* (12), 1951–1959. https://doi.org/10.1016/j.bbadis.2012.07.001

Vanneste, E., Voet, T., Le Caignec, C., Ampe, M., Konings, P., Melotte, C., Debrock, S., Amyere, M., Vikkula, M., Schuit, F., Fryns, J. P., Verbeke, G., D'Hooghe, T., Moreau, Y., & Vermeesch, J. R. (2009). Chromosome instability is common in human cleavage-stage embryos. *Nature Medicine, 15*(5), 577–583. https://doi.org/10.1038/nm.1924

Wang, H., Lin, X., Lyu, G., He, S., Dong, B., & Yang, Y. (2023). Chromosomal abnormalities in fetuses with congenital heart disease: A meta-analysis. *Archives of Gynecology and Obstetrics, 308*(3), 797–811. https://doi.org/10.1007/s00404-023-06910-3

Wang, X., & Zhang, G. (2025). The mitochondrial integrated stress response: A novel approach to anti-aging and pro-longevity. *Ageing Research Reviews, 103*, Article 102603. https://doi.org/10.1016/j.arr.2024.102603

Wen, H., Deng, H., Li, B., Chen, J., Zhu, J., Zhang, X., Yoshida, S., & Zhou, Y. (2025). Mitochondrial diseases: From molecular mechanisms to therapeutic advances. *Signal Transduction and Targeted Therapy, 10*(1). https://doi.org/10.1038/s41392-024-02044-3. Available from: https://www.nature.com/sigtrans

Wojcik, M. H., Schwartz, T. S., Yamin, I., Edward, H. L., Genetti, C. A., Towne, M. C., & Agrawal, P. B. (2018). Genetic disorders and mortality in infancy and early childhood: Delayed diagnoses and missed opportunities. *Genetics in Medicine, 20*(11), 1396–1404. https://doi.org/10.1038/gim.2018.17

Wong, C. (2023). UK first to approve CRISPR treatment for diseases: What you need to know. *Nature, 623*(7988), 676–677. https://doi.org/10.1038/d41586-023-03590-6

Xue, H., Yu, A., Lin, M., Chen, X., Guo, Q., Xu, L., & Huang, H. (2022). Efficiency of expanded noninvasive prenatal testing in the detection of fetal subchromosomal microdeletion and microduplication in a cohort of 31,256 single pregnancies. *Scientific Reports, 12*(1). https://doi.org/10.1038/s41598-022-24337-9. Available from: https://www.nature.com/srep/

Zafra, M. P., Schatoff, E. M., Katti, A., Foronda, M., Breinig, M., Schweitzer, A. Y., Simon, A., Han, T., Goswami, S., Montgomery, E., Thibado, J., Kastenhuber, E. R., Sánchez-Rivera, F. J., Shi, J., Vakoc, C. R., Lowe, S. W., Tschaharganeh, D. F., & Dow, L. E. (2018). Optimized base editors enable efficient editing in cells, organoids and mice. *Nature Biotechnology, 36*(9), 888–896. https://doi.org/10.1038/nbt.4194

Zhang, B., Zhou, Q., Chen, Y., Shi, Y., Zheng, F., Liu, J., & Yu, B. (2020). High false-positive non-invasive prenatal screening results for sex chromosome abnormalities: Are maternal factors the culprit? *Prenatal Diagnosis, 40*(4), 463–469. https://doi.org/10.1002/pd.5529. Available from: http://obgyn.onlinelibrary.wiley.com/hub/journal/10.1002/(ISSN)1097-0223/

Zuccaro, M. V., Xu, J., Mitchell, C., Marin, D., Zimmerman, R., Rana, B., Weinstein, E., King, R. T., Palmerola, K. L., Smith, M. E., Tsang, S. H., Goland, R., Jasin, M., Lobo, R., Treff, N., & Egli, D. (2020). Allele-specific chromosome removal after Cas9 cleavage in human embryos. *Cell, 183*(6), 1650. https://doi.org/10.1016/j.cell.2020.10.025

CHAPTER

Emergence of non-invasive prenatal screening in clinical practice

2

Riyaz Ahmad Rather

Department of Bioinformatics, Faculty of Engineering & Technology, Marwadi University, Rajkot, Gujarat, India

1. Introduction

The discovery of cell-free fetal DNA (cffDNA) in maternal plasma in 1997 marked a landmark milestone in reproductive genetics, laying the foundation for non-invasive prenatal screening (NIPS) (Lo et al., 1997). Subsequent advances in sequencing and bioinformatics laid concrete foundation for reliable detection and quantification of fetal-derived DNA fragments from a simple maternal blood sample, transforming prenatal screening paradigms (Chiu et al., 2008; Fan et al., 2008). NIPS is interchangeably referred as non-invasive prenatal testing (NIPT) as well by clinicians/geneticists and other reproductive biologists. The method has a wider scope in assessing the risk of common fetal aneuploidies, including trisomy 21, trisomy 18, and trisomy 13, as well as sex chromosome aneuploidies and selected microdeletion syndromes (Bianchi et al., 2014; Gil et al., 2017).

Traditionally, prenatal screening has been conducted using maternal age, biochemical markers, and ultrasonographic findings. These screening methods estimate the risk of fetal aneuploidies and selected genetic conditions. While these conventional approaches have contributed substantially to early risk stratification, they are limited by moderate sensitivity and specificity. Most importantly often need confirmatory invasive diagnostic procedures such as chorionic villus sampling (CVS) or amniocentesis. The later methods carry a small but measurable risk of procedure-related pregnancy loss (ACOG, 2016; Norton et al., 2012). The clinical need for safer, more accurate, and earlier screening strategies has therefore been a key to innovation in prenatal screening.

NIPS demonstrates substantially higher detection rates and lower false-positive rates, particularly for trisomy 21, where sensitivities exceeding 99% have been consistently reported (Gil et al., 2015; Taylor-Phillips et al., 2016). These performance characteristics have placed NIPS as a transformative technology that bridges the gap between screening and diagnostic testing, reducing the number of women

Non-Invasive Prenatal Screening of Rare Fetal Genetic Diseases. https://doi.org/10.1016/B978-0-443-27660-6.00008-0

referred for invasive procedures while maintaining high clinical confidence in risk assessment.

The rapid clinical uptake of NIPS reflects broader shifts toward precision medicine and patient-centered care in obstetrics. The ability to obtain highly accurate genetic risk information early in pregnancy facilitates timely counseling and individualized pregnancy management (Allyse et al., 2015). This chapter explores the emergence of NIPS as a foundation of modern prenatal care. The chapter traces the historical development of NIPS technologies, examines current methodologies and clinical applications, and discusses their integration into routine practice. The chapter further addresses benefits, limitations, ethical considerations, and future directions, providing a comprehensive foundation for understanding the role of NIPS in contemporary and evolving prenatal medicine.

2. Scientific basis of non-invasive prenatal screening

Cell-free DNA (cfDNA) originates primarily from apoptotic cells and is detectable in the bloodstream as short fragments, typically 140–170 base pairs in length (Fig. 2.1). During pregnancy, a proportion of this cfDNA is derived from the placenta and reflects the fetal genome; this fraction is referred to as cell-free fetal DNA (cffDNA) (Bianchi & Wilkins-Haug, 2014; Lo et al., 1997). Importantly, cffDNA can be detected as early as 5–7 weeks of gestation, with concentrations increasing progressively throughout pregnancy, enabling early risk assessment for fetal chromosomal abnormalities (Alberry et al., 2007; Wright & Burton, 2009). NIPS accuracy is dependent on the fetal fraction; defined as the proportion of total cfDNA in maternal plasma that is of fetal origin. Most clinical NIPS platforms require a minimum fetal fraction of approximately 4% to provide reliable detection of aneuploidy (Chiu et al., 2008; Palomaki et al., 2011). Fetal fraction is influenced by several biological and technical factors, including gestational age, maternal body mass index, placental health, and sample handling. Low fetal fraction is a recognized cause of test failure and may itself be associated with adverse pregnancy outcomes such as placental insufficiency or aneuploidy (Ashoor et al., 2013; Pergament et al., 2014).

The placental origin of cffDNA is a critical biological consideration in interpreting NIPS results. Because cffDNA predominantly reflects trophoblastic tissue rather than the fetus itself, discordant results may arise in cases of confined placental mosaicism, maternal chromosomal abnormalities, or maternal malignancy (Bianchi et al., 2015; Grati et al., 2014). Consequently, NIPS is classified as a high-performance screening test rather than a definitive diagnostic tool, and positive results require confirmation through invasive diagnostic procedures.

3. Development of non-invasive prenatal screening

During the late 1990s and early 2000s, efforts focused primarily on exploiting cffDNA for fetal sex determination and RhD genotyping in Rh-negative mothers.

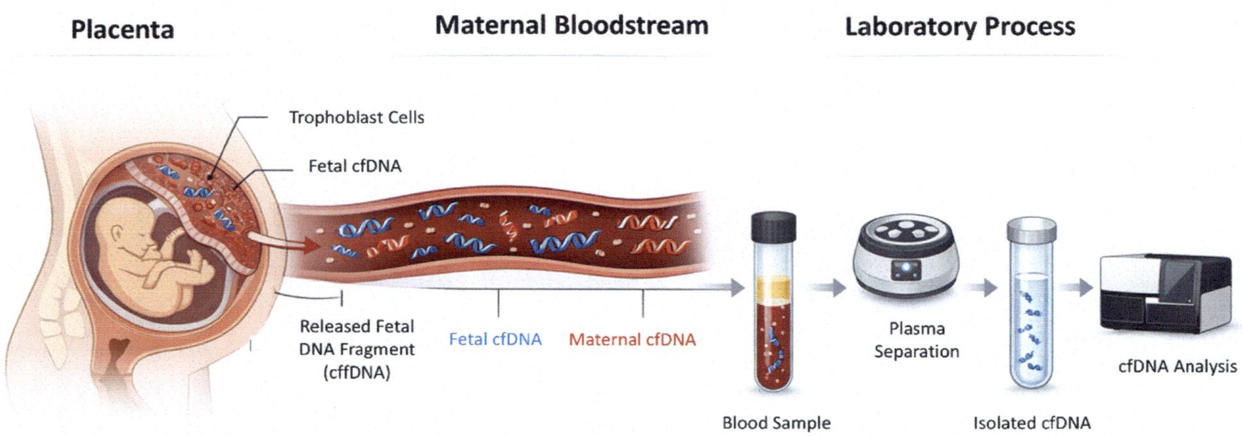

FIGURE 2.1

Biological basis of cell-free fetal DNA.

These early clinical applications demonstrated high accuracy and clinical utility, particularly in preventing unnecessary anti-D immunoglobulin administration (Lo et al., 1998; Mustafa et al., 2025). However, extending cffDNA analysis to detect chromosomal aneuploidies posed significant technical challenges, as fetal DNA represents only a small fraction of total circulating cfDNA and is mixed with a large background of maternal DNA. A major breakthrough occurred with the advent of massively parallel sequencing (MPS), also referred to as next-generation sequencing (NGS). In 2008, two independent groups demonstrated that shotgun sequencing of maternal plasma DNA followed by quantitative chromosomal counting could accurately detect fetal trisomy 21 (Chiu et al., 2008; Fan et al., 2008). These proof-of-concept studies established the feasibility of genome-wide cfDNA analysis for aneuploidy detection and laid the groundwork for clinical translation. Table 2.1 presents some key milestones in the development of the NIPS.

Subsequent large-scale validation studies in high-risk pregnancies confirmed the high sensitivity and specificity of NIPS for common autosomal trisomies, particularly trisomy 21 and trisomy 18 (Norton et al., 2012; Palomaki et al., 2011). These findings prompted rapid commercialization, with the first clinical NIPS assays becoming available in 2011. Early clinical adoption was primarily targeted toward women at increased risk for aneuploidy, including those of advanced maternal age or with abnormal conventional screening results. Technological evolution soon expanded beyond whole-genome shotgun sequencing. Targeted sequencing approaches reduced sequencing depth requirements and costs while maintaining analytical performance (Sparks et al., 2012). Single-nucleotide polymorphism

Table 2.1 Milestones in the development and clinical translation of NIPS.

Year	Milestone	Key advancement	Clinical significance
1997	Discovery of cffDNA in maternal plasma	Identification of fetal DNA fragments in maternal blood	Established biological basis for non-invasive testing
2007	Digital PCR for aneuploidy detection	Quantitative chromosome counting	Proof-of-concept for molecular counting
2008	Massively parallel sequencing introduced	Genome-wide cfDNA sequencing	Enabled reliable trisomy detection
2011	First commercial NIPS tests	Clinical availability	Transition from research to practice
2013–15	Large multicenter validation trials	High sensitivity/specificity confirmed	Expanded clinical confidence
2017–present	Population-level screening programs	First-tier or contingent NIPS	Mainstream integration

(SNP)-based methods were developed, enabling more sophisticated modeling of maternal and fetal haplotypes and improving detection of complex scenarios such as triploidy, vanishing twins, and some forms of mosaicism (Nicolaides et al., 2013; Rabinowitz et al., 2012). Digital PCR-based strategies also contributed to early assay development. As evidence accumulated, professional societies began to issue guidance on the clinical use of NIPS. Initially positioned as a screening option for high-risk pregnancies, NIPS was gradually endorsed for use in average-risk populations based on the performance data (Bianchi et al., 2014; Rose et al., 2022). This transition marked a critical phase in the maturation of NIPS from a niche technology to a mainstream component of prenatal care.

The scope of NIPS expanded over time to include sex chromosome aneuploidies and selected microdeletion syndromes, such as 22q11.2 deletion syndrome. However, performance for these additional conditions has generally been more variable than for common trisomies, prompting ongoing debate regarding appropriate test indications and counseling practices (Blagowidow et al., 2023; Taylor-Phillips et al., 2016). More recently, genome-wide cfDNA screening for large copy-number variants and rare autosomal aneuploidies has been introduced, further extending the technological reach of NIPS while simultaneously increasing interpretive complexity.

4. Methodologies in NIPS

NIPS includes a set of molecular and computational methodologies that analyze cfDNA in maternal plasma and infer fetal chromosomal status. Although these approaches differ in technical implementation and the target anomaly screening, they share a common workflow: maternal blood collection, plasma separation, cfDNA extraction, library preparation, and high-throughput molecular analysis. Continuous methodological refinement has yielded high analytical performance and broad clinical adoption.

4.1 Sample collection and cfDNA isolation

Maternal peripheral blood collected after 9–10 weeks of gestation is processed to separate plasma and minimize contamination with maternal genomic DNA released from leukocyte lysis. Specialized blood collection tubes containing cell-stabilizing agents are used to preserve cfDNA integrity during transport and storage (Bianchi & Wilkins-Haug, 2014; Norton et al., 2012). cfDNA is then extracted using silica-based or magnetic bead-based methods optimized for short DNA fragments. The resulting cfDNA pool contains a mixture of maternal and placental (fetal) DNA, with fragment sizes predominantly around 150 bp.

4.2 Massively parallel shotgun sequencing

Massively parallel shotgun sequencing (MPSS), also known as whole-genome sequencing (WGS), was the first widely validated NIPS methodology. In this approach, millions of cfDNA fragments are sequenced across the genome and aligned to a reference sequence. The proportion of reads mapping to each chromosome is quantified and compared with reference euploid distributions. A statistically significant overrepresentation of reads from a specific chromosome indicates a potential aneuploidy (Douville et al., 2020; Orr et al., 2015). MPSS provides genome-wide coverage and permits detection of common autosomal trisomies as well as sex chromosome aneuploidies. However, it requires relatively high sequencing depth and computational resources, contributing to higher costs compared with targeted approaches.

4.3 Targeted sequencing approaches

Targeted sequencing selectively analyzes predefined genomic regions or chromosomes of interest, such as chromosomes 13, 18, 21, X, and Y. By restricting sequencing to clinically relevant regions, this method reduces the total number of required reads while maintaining analytical sensitivity (Sparks et al., 2012). Chromosomal dosage is inferred by counting reads from targeted loci and comparing them with reference datasets. Targeted sequencing has become widely used in clinical laboratories because it offers a favorable balance between performance, cost, and throughput.

4.4 SNP-based NIPS

SNP-based methodologies use genotyping of thousands of polymorphic loci across the genome. These data are analyzed using probabilistic models integrating maternal genotype, fetal fraction, and expected inheritance patterns to determine fetal ploidy status (Caroselli et al., 2023; Nicolaides et al., 2013). SNP-based NIPS can distinguish between maternal and fetal DNA contributions and is particularly advantageous in complex scenarios, including triploidy, vanishing twin pregnancies, and some forms of mosaicism. Additionally, this approach can identify cases of consanguinity and detect maternal chromosomal abnormalities that may confound interpretation.

4.5 Digital PCR-based methods

Digital PCR (dPCR) was among the earliest quantitative approaches explored for NIPS. It partitions cfDNA into thousands of micro-reactions, providing precise counting of chromosome-specific targets. Differences in target copy number are used to infer aneuploidy (Lo et al., 2007).

4.6 Quality control and test reporting

Quality control increase the reliability of the test or screening methods. Quality metrics included in NIPS methods include fetal fraction thresholds, sequencing depth, library complexity, and noise parameters. Samples failing to meet predefined criteria are typically reported as "no-call" and may require repeat sampling. Results are expressed as high-risk or low-risk classifications rather than definitive diagnoses.

5. Integration into clinical practice

The integration of NIPS into routine clinical care represents one of the most significant paradigms shifts in modern obstetrics. NIPS introduces genomic information as a central component of prenatal risk assessment. Integration is therefore reconfiguration of counseling frameworks, clinical decision-making pathways, laboratory–clinic interfaces, and health-system policies. Effective integration requires alignment between scientific evidence, professional guidelines, infrastructure, reimbursement mechanisms, and clinician training.

5.1 Clinical positioning of NIPS in prenatal care pathways

Initially, NIPS was introduced as a secondary screening option for women considered at high risk for fetal aneuploidy. Over the last decade, robust prospective studies and meta-analyses have demonstrated that the analytical performance of NIPS is consistently high across both high-risk and average-risk populations, particularly for trisomy 21, trisomy 18, and trisomy 13. These data have driven a progressive shift toward offering NIPS as a universal or first-tier screening option (Fig. 2.2). Professional guidance from bodies such as American College of Obstetricians and Gynecologists (ACOG) and International Society for Prenatal Diagnosis (ISPD) supports offering NIPS to all pregnant individuals, irrespective of maternal age or baseline risk, provided that patients receive appropriate pretest counseling and have access to confirmatory diagnostic testing (ACOG, 2016; Benn et al., 2012). Clinically, this has resulted in three dominant models of positioning:

1. **First-tier screening**, where NIPS is the primary screening test offered at the first prenatal visit.
2. **Contingent screening**, where NIPS is offered after a positive or intermediate-risk result from conventional screening.
3. **Targeted screening**, where NIPS is reserved for predefined high-risk groups.

From a practical standpoint, most healthcare systems now offer NIPS from approximately 9–10 weeks' gestation, allowing risk assessment within the first trimester. This early positioning supports timely diagnostic confirmation, facilitates early pregnancy decision-making, and improves psychological preparedness for expectant parents.

5.2 Disease areas routinely integrated into NIPS

Clinical integration has been most firmly established for common autosomal trisomies. Trisomy 21 (Down syndrome), trisomy 18 (Edwards syndrome), and trisomy 13 (Patau syndrome) form the core of essentially all NIPS programs

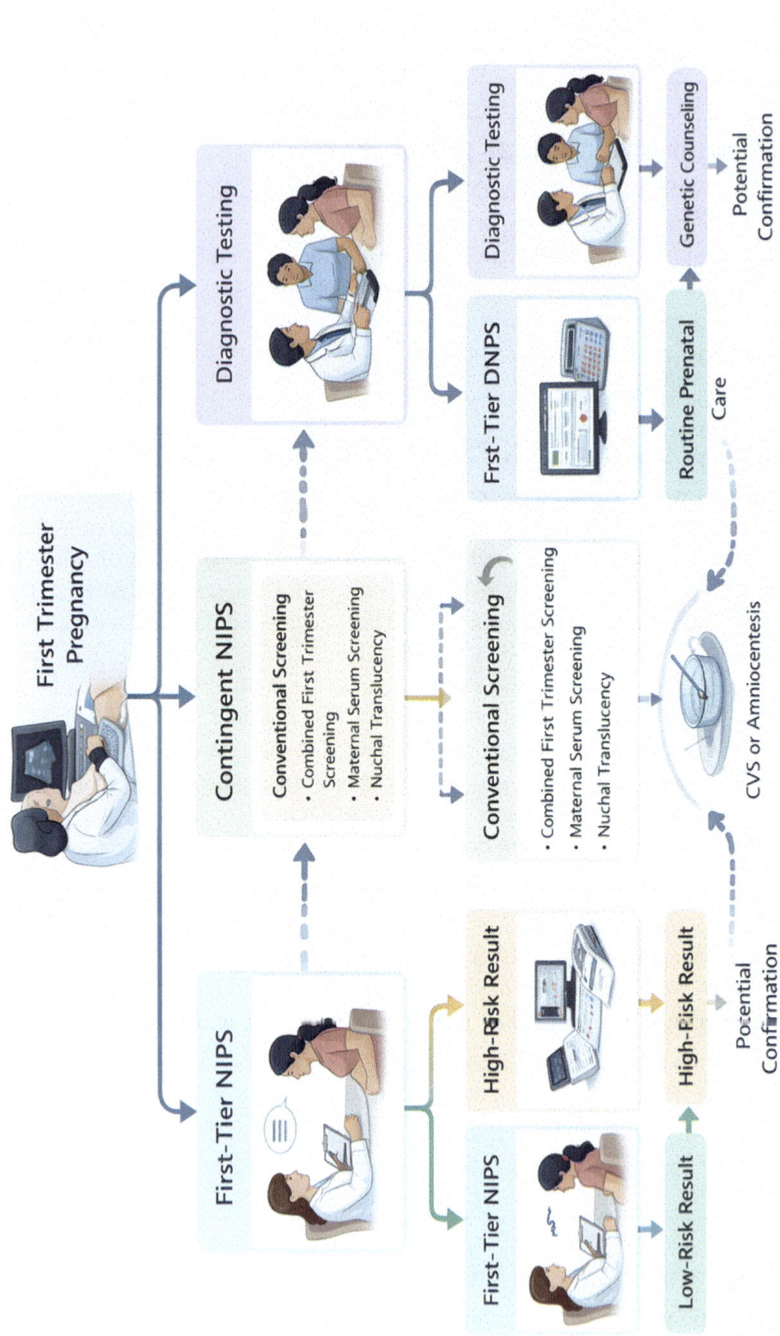

FIGURE 2.2

Integration pathway for NIPS in prenatal care.

worldwide. These conditions are characterized by relatively high prevalence, severe clinical impact, and well-validated NIPS performance metrics. Many laboratories and health systems also integrate screening for sex chromosome aneuploidies, including monosomy X, 47,XXY, 47,XXX, and 47,XYY (Rogol, 2023). However, clinical uptake of these results varies, as phenotypic severity is highly variable and counseling is often more complex. In contrast, microdeletion syndromes (such as 22q11.2 deletion syndrome) and rare autosomal trisomies are integrated inconsistently (Grati & Gross, 2019). Some private-sector laboratories include these conditions as part of expanded panels, whereas most publicly funded programs restrict routine screening to the three major trisomies. This cautious approach reflects lower prevalence, lower positive predictive values, and limited evidence that population-level screening improves clinical outcomes for these conditions. Consequently, clinicians must be prepared to explain differences between core screening targets and optional expanded panels and to emphasize that evidence strength is not uniform across all reported conditions. Table 2.2 presents disease categories integrated into NIPS.

5.3 Standard clinical workflow

NIPS is embedded within a standardized workflow designed to preserve its role as a screening test while maximizing clinical utility (Fig. 2.3). The pathway starts with pretest counseling, during which patients are informed about the purpose of NIPS, conditions screened, test accuracy, limitations, potential outcomes, and the possibility of inconclusive results (Hsieh et al., 2025). Counseling also emphasizes that NIPS does not replace diagnostic testing. Following maternal blood collection and laboratory analysis, results are reported as high risk or low risk for specific conditions, accompanied by fetal fraction and quality metrics. Low-risk results generally lead to continuation of routine prenatal care. High-risk results trigger referral for confirmatory diagnostic testing using chorionic villus sampling or amniocentesis, along with genetic counseling to discuss implications, prognosis, and options (Dungan et al., 2023). This integrated workflow has become widely accepted because it reduces unnecessary invasive procedures, concentrates diagnostic testing in pregnancies with the highest likelihood of aneuploidy. The workflow also streamlines referral patterns within maternal–fetal medicine services. The proposed workflow is presented in Fig. 2.3.

5.4 Country-wise implementation models

United States: NIPS is extensively integrated as both first-tier and patient-choice screening in public and private healthcare settings (Agarwal et al., 2013; Connor et al., 2025). Most obstetric practices offer NIPS to all pregnant patients early in pregnancy. Integration has been supported by insurance coverage expansion and strong laboratory infrastructure. Clinically, this has led to widespread replacement of conventional serum screening as the primary aneuploidy risk assessment tool in many practices (Oyelese et al., 2025).

Table 2.2 Disease categories integrated into nips and strength of evidence.

Disease category	Example conditions	Typical sensitivity	Typical PPV	Level of clinical validation	Routine integration status
Common autosomal trisomies	Trisomy 21, 18, 13	>99% (T21)	High	Very strong	Universal
Sex chromosome aneuploidies	45,X; 47,XXY	90%–95%	Moderate	Strong	Common
Microdeletions	22q11.2 deletion	Variable	Low–moderate	Limited	Selective
Rare autosomal trisomies	RATs	Variable	Low	Emerging	Limited
Genome-wide CNVs	>7–10 Mb	Variable	Low	Emerging	Limited

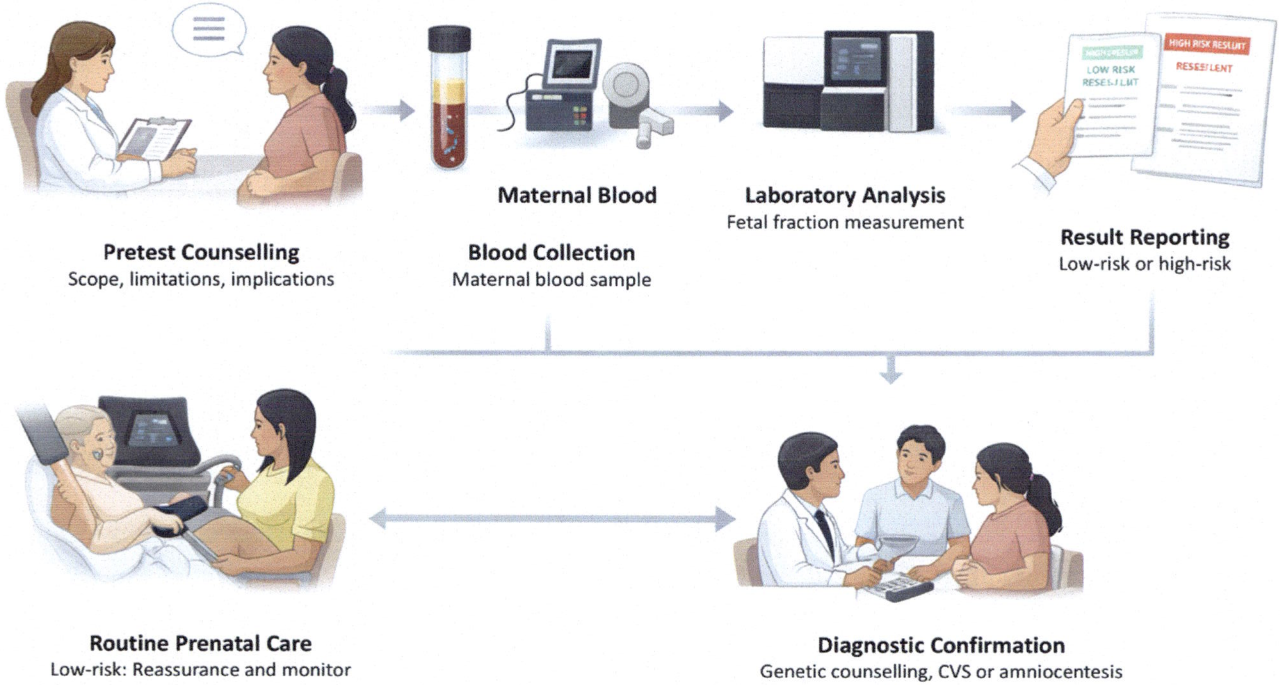

FIGURE 2.3

NIPS clinical workflow.

United Kingdom: Within NHS England, NIPS is integrated as a publicly funded contingent screening test. Women with an elevated risk on combined first-trimester screening are offered NIPS for trisomies 21, 18, and 13 (Takahashi et al., 2022). This model has achieved high detection rates while maintaining cost-effectiveness and equitable access. Importantly, standardized national information leaflets and consent processes ensure consistency of counseling across the country.

Netherlands: The Netherlands has implemented nationwide first-tier NIPS through a government-coordinated program that allows all pregnant individuals to choose NIPS as their primary screening test (Lannoo et al., 2023). Core screening targets include trisomies 21, 18, and 13, with optional reporting of additional findings under clearly defined consent categories. Continuous national data collection supports quality monitoring and policy refinement.

Australia: NIPS is widely available, primarily through self-funded testing, and commonly used as primary screening for common trisomies (Hui & Halliday, 2023). Professional guidance supports its use across all risk groups. Integration has been particularly strong among women aged 35 years and older, although uptake is increasing in average-risk populations.

China: China has experienced rapid large-scale adoption of NIPS as first-tier screening in many tertiary hospitals and commercial laboratories (Yang et al., 2022). Screening for trisomies 21, 18, and 13 is universal, and sex chromosome aneuploidies are frequently included. Variability in counseling practices and regulatory oversight remains a challenge, underscoring the need for national standardization.

India: In India, NIPS is mainly integrated within private-sector prenatal care and used predominantly for high-risk pregnancies. Routine screening focuses on trisomies 21, 18, and 13. Integration occurs alongside strict legal regulations prohibiting fetal sex disclosure, necessitating careful documentation and compliance (Ravitsky et al., 2021; Verma, 2014). Country-wise models for clinical integration of NIPS is presented in Table 2.3.

6. Ethical, legal, and social implications

The rapid clinical adoption of NIPS has generated profound ethical, legal, and social implications that extend beyond laboratory performance and clinical utility. Because NIPS provides early and increasingly comprehensive genetic information about the fetus through a simple maternal blood test, it reshapes how reproductive choices are made. Ethical integration of NIPS, therefore, requires careful attention to informed consent, reproductive autonomy, equity, regulatory oversight, and the broader societal consequences of expanding prenatal genetic knowledge.

6.1 Informed consent and reproductive autonomy

A key ethical principle in NIPS implementation is respect for patient autonomy. Although the test is technically simple, its implications are complex. Patients

Table 2.3 Country-wise models for clinical integration of non-invasive prenatal screening.

Country/ Region	Funding model	Positioning of NIPS	Core conditions screened	Population coverage
United States	Insurance/ private	First-tier or choice-based	T21, T18, T13 ± SCA	Universal offer
United Kingdom	Public (NHS)	Contingent	T21, T18, T13	Screen-positive women
Netherlands	Public	First-tier	T21, T18, T13	Universal offer
Australia	Self-funded	First-tier	T21, T18, T13	Broad uptake
China	Mixed	First-tier	T21, T18, T13 ± SCA	Large-scale
India	Private	Targeted	T21, T18, T13	High-risk

must understand that NIPS is a screening test rather than a diagnostic test, that results are probabilistic, and that positive findings require confirmatory invasive testing (Dondorp et al., 2015). Inadequate counseling risks transforming NIPS into a routine blood test undertaken without meaningful reflection, thereby undermining informed choice. ACOG and ISPD emphasize that pretest counseling should cover test scope, detection rates, false-positive and false-negative possibilities, limitations, and potential outcomes. From an ethical perspective, consent must be voluntary, informed, and non-directive (Kater-Kuipers et al., 2020). Clinicians should avoid framing NIPS as something that "should" be done, instead presenting it as an option aligned with the patient's values, beliefs, and preferences. At the same time, NIPS enhances reproductive autonomy by enabling earlier access to high-quality information. Earlier detection of high-risk pregnancies allows more time for confirmatory testing, reflection, and decision-making. However, this empowerment coexists with the potential for subtle pressure to test or to act on results in particular ways, especially in healthcare cultures that prioritize risk avoidance.

6.2 Screening versus diagnosis: Ethical boundaries

Ethically, it is essential to maintain a clear boundary between screening and diagnosis. Because NIPS has very high sensitivity for common trisomies, patients may mistakenly interpret results as definitive (Bowman-Smart et al., 2024). Overstating certainty can lead to irreversible decisions, including pregnancy termination, based solely on screening results. Ethical practice therefore, requires consistent reinforcement that diagnostic confirmation is mandatory before irreversible clinical actions. Another related issue is the expansion of NIPS panels beyond well-validated conditions. While screening for trisomy 21, 18, and 13 is supported by strong evidence, performance for microdeletions and rare copy-number variants is more variable. Offering such tests without clearly communicating uncertainty may expose patients to anxiety, unnecessary invasive procedures, and ambiguous findings. Ethically responsible integration requires proportionality: the strength of evidence should match the seriousness of clinical decisions that may follow.

6.3 Disability rights and social perceptions

One of the most debated social implications of NIPS concerns its potential impact on individuals and communities living with genetic conditions, particularly Down syndrome (Kellogg et al., 2014). Critics argue that widespread screening may contribute to stigmatization. Advocates of disability rights emphasize that expanded prenatal screening should be accompanied by balanced, accurate information about living with these conditions, including positive aspects and available supports. From a clinical standpoint, counseling should avoid presenting conditions such as Down syndrome exclusively in terms of deficits (Zerres et al., 2021). Ethical counseling acknowledges medical challenges while also recognizing variability in quality of life, educational achievement, and social participation. The goal is not to steer decisions but to support informed, values-consistent choices.

6.4 Equity of access and health-system justice

Equity is a major ethical concern in NIPS implementation. In high-income countries, NIPS is increasingly publicly funded or reimbursed, whereas in many low- and middle-income countries it remains accessible primarily through private payment (Löwy, 2022). This creates a two-tier system in which wealthier individuals benefit from more accurate screening and reduced exposure to invasive procedures, while disadvantaged populations rely on less accurate methods. Publicly funded contingent or first-tier models, such as those implemented within NHS England and national programs in the Netherlands, represent attempts to balance cost-effectiveness with equitable access (Bilardo, 2021). In contrast, settings where NIPS is predominantly self-funded may inadvertently widen existing disparities in prenatal care. Ethically, health systems must consider whether access to high-quality aneuploidy screening constitutes a basic component of prenatal care and, if so, how to provide it fairly across populations.

6.5 Legal and regulatory frameworks

Legal oversight of NIPS varies widely by country. In many jurisdictions, NIPS is regulated as a laboratory-developed test, with oversight focused on analytical validity rather than clinical outcomes (Rather & Saha, 2023). Other countries integrate NIPS into national screening programs with standardized protocols, reporting formats, and quality assurance mechanisms. Regulation also intersects with reproductive law. In countries where termination of pregnancy is legally restricted or gestational-age limited, early access to NIPS has particular significance. Earlier detection may expand practical access to lawful options, whereas delays or lack of access may constrain reproductive choice. In some countries, including India,

NIPS exists alongside strict legal prohibitions on fetal sex determination (Verma, 2003). Clinicians must navigate these laws carefully, ensuring that NIPS is not used for sex selection and that documentation and reporting comply with regulatory requirements.

6.6 Incidental and secondary findings

NIPS occasionally reveals information unrelated to the primary screening target, such as suspected maternal chromosomal abnormalities, confined placental mosaicism, or rarely, patterns suggestive of maternal malignancy (Evans et al., 2016). These findings raise ethical questions about disclosure: what should be reported, how certain must the finding be, and what follow-up should be offered? Ethically sound practice requires predefined laboratory policies regarding incidental findings and clear communication to patients during pretest counseling about the possibility of such results. Clinicians must be prepared to coordinate appropriate referrals while minimizing unnecessary alarms.

6.7 Commercialization and marketing ethics

The rapid commercialization of NIPS has introduced competitive marketing, direct-to-consumer advertising, and expanded test panels promoted as technological progress (Agarwal et al., 2013; Löwy, 2022). Ethical concerns arise when marketing emphasizes breadth rather than clinical validity, potentially overstating benefits and minimizing limitations. Clinicians play a crucial role as ethical gatekeepers, helping patients interpret promotional claims, focusing discussions on evidence-based indications, and resisting pressure to order expanded testing without clear clinical justification.

7. Challenges and limitations

Although NIPS has substantially improved prenatal risk assessment, it has important inherent limitations that must be recognized to prevent misinterpretation and inappropriate clinical decision-making. A key challenge arises from the biological origin of cell-free fetal DNA, which is derived predominantly from placental trophoblasts rather than directly from fetal tissues. Consequently, discrepancies between placental and fetal karyotypes, known as confined placental mosaicism, can lead to false-positive or false-negative NIPS results. Maternal chromosomal mosaicism or copy-number variants may further confound analysis, and rare cases of abnormal cfDNA patterns associated with maternal malignancy have been reported. These biological factors impose a fundamental ceiling on achievable accuracy and explain why NIPS must remain a screening rather than a diagnostic test.

Analytical performance is also influenced by the fetal fraction, the proportion of fetal DNA within total circulating cfDNA. Low fetal fraction, which is more common at early gestational ages, in individuals with higher body mass index, or in certain placental disorders, increases the likelihood of test failure or inconclusive ("no-call") results. Repeat sampling may be required, delaying risk assessment and increasing patient anxiety. Importantly, low fetal fraction itself has been associated with an increased risk of aneuploidy and adverse pregnancy outcomes, creating a clinical dilemma in which higher-risk pregnancies may yield less informative results.

Another major limitation relates to the scope of conditions that can be reliably detected. Strongest evidence exists for trisomy 21, trisomy 18, and trisomy 13, whereas performance for sex chromosome aneuploidies is generally lower and more variable. Screening for microdeletion syndromes, rare autosomal trisomies, and genome-wide copy-number variants shows substantially lower positive predictive values, particularly in low-risk populations. Despite the commercial expansion of test panels, many single-gene disorders, balanced chromosomal rearrangements, and small pathogenic variants are not reliably detectable by standard NIPS. This gap between perceived and actual test capability can lead to false reassurance or unnecessary concern.

Predictive values of NIPS are highly dependent on disease prevalence. In populations with low baseline risk, even highly specific tests may yield a considerable proportion of false-positive results for rare conditions. Conversely, false-negative results, although uncommon for common trisomies, do occur and may contribute to missed diagnoses if clinicians or patients overestimate the comprehensiveness of NIPS. Therefore, normal NIPS results do not eliminate the need for ongoing ultrasound surveillance and clinical assessment.

Interpretation and counseling present additional challenges. The probabilistic nature of NIPS results, variability in reporting formats between laboratories, and increasing availability of expanded panels complicate communication with patients. Inadequate counseling may result in misunderstanding of risk, unrealistic expectations, or premature, irreversible decisions. Clinicians must therefore devote sufficient time to explaining test limitations, emphasizing the need for confirmatory diagnostic testing after positive results, and contextualizing findings within the broader clinical picture.

Finally, system-level challenges influence the effectiveness of NIPS integration. Cost, reimbursement policies, unequal access between public and private sectors, and variability in laboratory quality assurance can create disparities in who benefits from NIPS. Rapid technological evolution also imposes ongoing educational demands on healthcare providers, who must remain current with evolving capabilities and evidence.

8. Conclusion

NIPS has fundamentally reshaped the landscape of prenatal care by enabling early, highly accurate, and safe assessment of fetal chromosomal abnormalities through analysis of cell-free fetal DNA in maternal blood. From its origins as a scientific discovery to its widespread clinical adoption, NIPS represents a successful translation of genomic technology into routine obstetric practice. The technology has demonstrated consistently superior performance compared with conventional biochemical screening, particularly for trisomies 21, 18, and 13, and has contributed to substantial reductions in invasive diagnostic procedures and procedure-related pregnancy loss. Beyond its technical strengths, the true value of NIPS lies in its integration within comprehensive prenatal care pathways that emphasize informed choice, appropriate counseling, and confirmatory diagnostic testing when indicated. When used responsibly, NIPS supports patient-centered care by providing timely, reliable risk information that facilitates individualized pregnancy management and shared decision-making.

At the same time, important challenges remain. Biological constraints, such as placental mosaicism and low fetal fraction, limit absolute accuracy, while expanding test panels introduce interpretive complexity and ethical concerns. Variability in access, regulation, and counseling practices across countries highlights the need for harmonized standards and equitable implementation strategies.

As prenatal medicine continues to evolve, NIPS should be viewed not as a replacement for existing diagnostic modalities, but as a complementary tool within a layered screening and diagnostic framework. Ongoing research, technological refinement, and clinician education will be essential to optimize performance, clarify appropriate indications, and address ethical, legal, and social implications. Ultimately, the responsible integration of NIPS has the potential to enhance prenatal care by combining scientific innovation with compassionate, informed, and equitable clinical practice.

References

Agarwal, A., Sayres, L. C., Cho, M. K., Cook-Deegan, R., & Chandrasekharan, S. (2013). Commercial landscape of noninvasive prenatal testing in the United States. *Prenatal Diagnosis, 33*(6), 521–531. https://doi.org/10.1002/pd.4101

Alberry, M., Maddocks, D., Jones, M., Abdel Hadi, M., Abdel-Fattah, S., Avent, N., & Soothill, P. W. (2007). Free fetal DNA in maternal plasma in anembryonic pregnancies: Confirmation that the origin is the trophoblast. *Prenatal Diagnosis, 27*(5), 415–418. https://doi.org/10.1002/pd.1700

Allyse, M., Minear, M. A., Berson, E., Sridhar, S., Rote, M., Hung, A., & Chandrasekharan, S. (2015). Non-invasive prenatal testing: A review of international implementation and challenges. *International Journal of Women's Health, 7*, 113–126. https://doi.org/10.2147/IJWH.S67124

American College of Obstetricians and Gynecologists (ACOG). (2016). Practice Bulletin No. 163: Screening for fetal aneuploidy. *Obstetrics & Gynecology, 127*(5), e123–e137.

Ashoor, G., Syngelaki, A., Wang, E., Struble, C., Oliphant, A., Song, K., & Nicolaides, K. H. (2013). Trisomy 13 detection in the first trimester of pregnancy using a chromosome-selective cell-free DNA analysis method. *Ultrasound in Obstetrics and Gynecology, 41*(1), 21–25. https://doi.org/10.1002/uog.12299

Benn, P., Borrell, A., Cuckle, H., Dugoff, L., Gross, S., Johnson, J. A., Maymon, R., Odibo, A., Schielen, P., Spencer, K., Wright, D., & Yaron, Y. (2012). Prenatal Detection of Down Syndrome using Massively Parallel Sequencing (MPS): A rapid response statement from a committee on behalf of the Board of the International Society for Prenatal Diagnosis, 24 October 2011. *Prenatal Diagnosis, 32*(1), 1–2. https://doi.org/10.1002/pd.2919

Bianchi, D. W., Chudova, D., Sehnert, A. J., Bhatt, S., Murray, K., Prosen, T. L., Garber, J. E., Wilkins-Haug, L., Vora, N. L., Warsof, S., Goldberg, J., Ziainia, T., & Halks-Miller, M. (2015). Noninvasive prenatal testing and incidental detection of occult maternal malignancies. *JAMA, the Journal of the American Medical Association, 314*(2), 162–169. https://doi.org/10.1001/jama.2015.7120

Bianchi, D. W., Lamar Parker, R., Wentworth, J., Madankumar, R., Saffer, C., Das, A. F., Craig, J. A., Chudova, D. I., Devers, P. L., Jones, K. W., Oliver, K., Rava, R. P., & Sehnert, A. J. (2014). DNA sequencing versus standard prenatal aneuploidy screening. *New England Journal of Medicine, 370*(9), 799–808. https://doi.org/10.1056/NEJMoa1311037

Bianchi, D. W., & Wilkins-Haug, L. (2014). Integration of noninvasive DNA testing for aneuploidy into prenatal care: What has happened since the rubber met the road? *Clinical Chemistry, 60*(1), 78–87. https://doi.org/10.1373/clinchem.2013.202663UnitedStates

Bilardo, C. M. (2021). The implementation of non-invasive prenatal testing (NIPT) in the Netherlands. *Journal of Perinatal Medicine, 49*(8), 941–944. https://doi.org/10.1515/jpm-2021-0290

Blagowidow, N., Nowakowska, B., Schindewolf, E., Grati, F. R., Putotto, C., Breckpot, J., Swillen, A., Crowley, T. B., Loo, J. C. Y., Lairson, L. A., Óskarsdóttir, S., Boot, E., Garcia-Minaur, S., Cristina Digilio, M., Marino, B., Coleman, B., Moldenhauer, J. S., Bassett, A. S., & McDonald-McGinn, D. M. (2023). Prenatal screening and diagnostic considerations for 22q11.2 microdeletions. *Genes, 14*(1). https://doi.org/10.3390/genes14010160

Bowman-Smart, H., Perrot, A., & Horn, R. (2024). Supporting patient decision-making in non-invasive prenatal testing: A comparative study of professional values and practices in England and France. *BMC Medical Ethics, 25*(1). https://doi.org/10.1186/s12910-024-01032-0

Caroselli, S., Figliuzzi, M., Picchetta, L., Cogo, F., Zambon, P., Pergher, I., Girardi, L., Patassini, C., Poli, M., Bakalova, D., Cimadomo, D., Findikli, N., Coban, O., Serdarogullari, M., Favero, F., Bortolato, S., Anastasi, A., Capodanno, F., Gallinelli, A., … Capalbo, A. (2023). Improved clinical utility of preimplantation genetic testing through the integration of ploidy and common pathogenic microdeletions analyses. *Human Reproduction, 38*(4), 762–775. https://doi.org/10.1093/humrep/dead033

Chiu, R. W. K., Chan, K. C. A., Gao, Y., Lau, V. Y. M., Zheng, W., Leung, T. Y., Foo, C. H. F., Xie, B., Tsui, N. B. Y., Lun, F. M. F., Zee, B. C. Y., Lau, T. K., Cantor, C. R., & Lo, Y. M. D. (2008). Noninvasive prenatal diagnosis of fetal chromosomal aneuploidy by massively parallel genomic sequencing of DNA in maternal plasma. *Proceedings of the National Academy of Sciences of the United States of America, 105*(51), 20458–20463. https://doi.org/10.1073/pnas.0810641105

Connor, C., Sato, T., Bianchi, D. W., Fenton, K., Somani, E., Turriff, A., Berkman, B., & Hendriks, S. (2025). Comparing the introduction and implementation of noninvasive prenatal testing (NIPT) in Japan, the Netherlands, and the United States: An integrative review. *Prenatal Diagnosis, 45*(10), 1244–1264. https://doi.org/10.1002/pd.6708

Dennis Lo, Y. M., Corbetta, N., Chamberlain, P. F., Rai, V., Sargent, I. L., Redman, C. W. G., & Wainscoat, J. S. (1997). Presence of fetal DNA in maternal plasma and serum. *Lancet, 350*(9076), 485–487. https://doi.org/10.1016/S0140-6736(97)02174-0

Dondorp, W., De Wert, G., Bombard, Y., Bianchi, D. W., Bergmann, C., Borry, P., Chitty, L. S., Fellmann, F., Forzano, F., Hall, A., Henneman, L., Howard, H. C., Lucassen, A., Ormond, K., Peterlin, B., Radojkovic, D., Rogowski, W., Soller, M., Tibben, A., … Cornel, M. C. (2015). Non-invasive prenatal testing for aneuploidy and beyond: Challenges of responsible innovation in prenatal screening. *European Journal of Human Genetics, 23*(11), 1438–1450. https://doi.org/10.1038/ejhg.2015.57

Douville, C., Cohen, J. D., Ptak, J., Popoli, M., Schaefer, J., Silliman, N., Dobbyn, L., Schoen, R. E., Tie, J., Gibbs, P., Goggins, M., Wolfgang, C. L., Wang, T.-L., Shih, I.-M., Karchin, R., Lennon, A. M., Hruban, R. H., Tomasetti, C., Bettegowda, C., … Vogelstein, B. (2020). Assessing aneuploidy with repetitive element sequencing. *Proceedings of the National Academy of Sciences, 117*(9), 4858–4863. https://doi.org/10.1073/pnas.1910041117

Dungan, J. S., Klugman, S., Darilek, S., Malinowski, J., Akkari, Y. M. N., Monaghan, K. G., Erwin, A., & Best, R. G. (2023). Noninvasive prenatal screening (NIPS) for fetal chromosome abnormalities in a general-risk population: An evidence-based clinical guideline of the American College of Medical Genetics and Genomics (ACMG). *Genetics in Medicine, 25*(2). https://doi.org/10.1016/j.gim.2022.11.004

Evans, M. I., Wapner, R. J., & Berkowitz, R. L. (2016). Noninvasive prenatal screening or advanced diagnostic testing: Caveat emptor. *American Journal of Obstetrics and Gynecology, 215*(3), 298–305. https://doi.org/10.1016/j.ajog.2016.04.029

Fan, H. C., Blumenfeld, Y. J., Chitkara, U., Hudgins, L., & Quake, S. R. (2008). Noninvasive diagnosis of fetal aneuploidy by shotgun sequencing DNA from maternal blood. *Proceedings of the National Academy of Sciences, 105*(42), 16266–16271. https://doi.org/10.1073/pnas.0808319105

Gil, M. M., Accurti, V., Santacruz, B., Plana, M. N., & Nicolaides, K. H. (2017). Analysis of cell-free DNA in maternal blood in screening for aneuploidies: Updated meta-analysis. *Ultrasound in Obstetrics and Gynecology, 50*(3), 302–314. https://doi.org/10.1002/uog.17484

Gil, M. M., Quezada, M. S., Revello, R., Akolekar, R., & Nicolaides, K. H. (2015). Analysis of cell-free DNA in maternal blood in screening for aneuploidies: Meta-analysis. *Fetal Diagnosis and Therapy, 37*(2), 82–91.

Grati, F. R., & Gross, S. J. (2019). Noninvasive screening by cell-free DNA for 22q11.2 deletion: Benefits, limitations, and challenges. *Prenatal Diagnosis, 39*(2), 70–80. https://doi.org/10.1002/pd.5391

Grati, F. R., Malvestiti, F., Ferreira, J. C. P. B., Bajaj, K., Gaetani, E., Agrati, C., Grimi, B., Dulcetti, F., Ruggeri, A. M., De Toffol, S., Maggi, F., Wapner, R., Gross, S., & Simoni, G. (2014). Fetoplacental mosaicism: Potential implications for false-positive and false-negative noninvasive prenatal screening results. *Genetics in Medicine, 16*(8), 620–624. https://doi.org/10.1038/gim.2014.3

Hsieh, V., Sherer, D. M., Davydovych, K., Kheyman, M., & Dalloul, M. (2025). The art (and science) of individualized selection of non-invasive prenatal screening (NIPS). *International Journal of Women's Health, 17*, 1271–1283. https://doi.org/10.2147/IJWH.S437214

Hui, L., & Halliday, J. (2023). A decade of non-invasive prenatal screening in Australia: National impact on prenatal screening and diagnostic testing. *The Australian and New Zealand Journal of Obstetrics and Gynaecology, 63*(2), 264–267. https://doi.org/10.1111/ajo.13638

Kater-Kuipers, A., de Beaufort, I. D., Galjaard, R. J. H., & Bunnik, E. M. (2020). Rethinking counselling in prenatal screening: An ethical analysis of informed consent in the context of non-invasive prenatal testing (NIPT). *Bioethics, 34*(7), 671–678. https://doi.org/10.1111/bioe.12760

Kellogg, G., Slattery, L., Hudgins, L., & Ormond, K. (2014). Attitudes of mothers of children with down syndrome towards noninvasive prenatal testing. *Journal of Genetic Counseling, 23*(5), 805–813. https://doi.org/10.1007/s10897-014-9694-7

Löwy, I. (2022). Non-invasive prenatal testing: A diagnostic innovation shaped by commercial interests and the regulation conundrum. *Social Science & Medicine, 304*, Article 113064. https://doi.org/10.1016/j.socscimed.2020.113064

Lannoo, L., van der Meij, K. R. M., Bekker, M. N., De Catte, L., Deckers, S., Devriendt, K., Roggen, N., Galjaard, R. J. H., Gitsels-van der Wal, J., Macville, M. V. E., Martin, L., Sistermans, E. A., Van Calsteren, K., Van Keirsbilck, J., Crombag, N., & Henneman, L. (2023). A cross-country comparison of pregnant women's decision-making and perspectives when opting for non-invasive prenatal testing in the Netherlands and Belgium. *Prenatal Diagnosis, 43*(3), 294–303. https://doi.org/10.1002/pd.6329

Lo, Y. M. D., Hjelm, N. M., Fidler, C., Sargent, I. L., Murphy, M. F., Chamberlain, P. F., Poon, P. M. K., Redman, C. W. G., & Wainscoat, J. S. (1998). Prenatal diagnosis of fetal RhD status by molecular analysis of maternal plasma. *New England Journal of Medicine, 339*(24), 1734–1738. https://doi.org/10.1056/NEJM199812103392402

Lo, Y. M. D., Lun, F. M. F., Allen Chan, K. C., Tsui, N. B. Y., Chong, K. C., Lau, T. K., Leung, T. Y., Zee, B. C. Y., Cantor, C. R., & Chiu, R. W. K. (2007). Digital PCR for the molecular detection of fetal chromosomal aneuploidy. *Proceedings of the National Academy of Sciences, 104*(32), 13116–13121. https://doi.org/10.1073/pnas.0705765104

Mustafa, H. J., Najjariasl, P., Aghajani, F., Sambatur, E. V., Khalil, A., Moise, K. J., & Shamshirsaz, A. A. (2025). Diagnostic accuracy of cell-free DNA for the determination of fetal red blood cell antigen genotype: A systematic review and meta-analysis. *American Journal of Obstetrics and Gynecology, 233*(5), 428. https://doi.org/10.1016/j.ajog.2025.05.004

Nicolaides, K. H., Syngelaki, A., Ashoor, G., Birdir, C., & Touzet, G. (2013). Noninvasive prenatal testing for fetal trisomies in a routinely screened first-trimester population. *Obstetrical and Gynecological Survey, 68*(3), 173–175. https://doi.org/10.1097/OGX.0b013e318285bf66

Norton, M. E., Brar, H., Weiss, J., Karimi, A., Laurent, L. C., Caughey, A. B., Rodriguez, M. H., Williams, J., Mitchell, M. E., Adair, C. D., Lee, H., Jacobsson, B., Tomlinson, M. W., Oepkes, D., Hollemon, D., Sparks, A. B., Oliphant, A., & Song, K. (2012). Non-Invasive Chromosomal Evaluation (NICE) Study: Results of a multicenter prospective cohort study for detection of fetal trisomy 21 and trisomy 18. *American Journal of Obstetrics and Gynecology, 207*(2), 137. https://doi.org/10.1016/j.ajog.2012.05.021

Orr, B., Godek, K. M., & Compton, D. (2015). Aneuploidy. *Current Biology, 25*(13), Article R538. https://doi.org/10.1016/j.cub.2015.05.010

Oyelese, Y., Schioppo, D., & O'Brien, B. (2025). Prenatal screening and diagnosis: Time for a paradigm shift. *American Journal of Perinatology, 42*(04), 538–545. https://doi.org/10.1055/a-2312-8824

Palomaki, G. E., Kloza, E. M., Lambert-Messerlian, G. M., Haddow, J. E., Neveux, L. M., Ehrich, M., Van Den Boom, D., Bombard, A. T., Deciu, C., Grody, W. W., Nelson, S. F., & Canick, J. A. (2011). DNA sequencing of maternal plasma to detect Down syndrome: An international clinical validation study. *Genetics in Medicine, 13* (11), 913–920. https://doi.org/10.1097/GIM.0b013e3182368a0e

Pergament, E., Cuckle, H., Zimmermann, B., Banjevic, M., Sigurjonsson, S., Ryan, A., Hall, M. P., Dodd, M., Lacroute, P., Stosic, M., Chopra, N., Hunkapiller, N., Prosen, D. E., McAdoo, S., Demko, Z., Siddiqui, A., Hill, M., & Rabinowitz, M. (2014). Single-nucleotide polymorphism–based noninvasive prenatal screening in a high-risk and low-risk cohort. *Obstetrics & Gynecology, 124*(2), 210–218. https://doi.org/10.1097/aog.0000000000000363

Rabinowitz, M., Gemelos, G., Banjevic, M., Ryan, A., Demko, Z., Hill, M., & Gross, S. J. (2012). *Methods for non-invasive prenatal testing of fetal aneuploidy.* US Patent, 20120003645.

Rather, R. A., & Saha, S. C. (2023). Reappraisal of evolving methods in non-invasive prenatal screening: Discovery, biology and clinical utility. *Heliyon, 9*(3). https://doi.org/10.1016/j.heliyon.2023.e13923

Ravitsky, V., Roy, M. C., Haidar, H., Henneman, L., Marshall, J., Newson, A. J., Ngan, O. M. Y., & Nov-Klaiman, T. (2021). The emergence and global spread of noninvasive prenatal testing. *Annual Review of Genomics and Human Genetics, 22*, 309–338.

Rogol, A. D. (2023). Sex chromosome aneuploidies and fertility: 47,XXY, 47,XYY, 47,XXX and 45,X/47,XXX. *Endocrine Connections, 12*(9). https://doi.org/10.1530/EC-22-0440

Rose, N. C., Barrie, E. S., Malinowski, J., Jenkins, G. P., McClain, M. R., LaGrave, D., & Leung, M. L. (2022). Systematic evidence-based review: The application of noninvasive prenatal screening using cell-free DNA in general-risk pregnancies. *Genetics in Medicine, 24*(7), 1379–1391. https://doi.org/10.1016/j.gim.2022.03.019

Sparks, A. B., Struble, C. A., Wang, E. T., Song, K., & Oliphant, A. (2012). Noninvasive prenatal detection and selective analysis of cell-free DNA obtained from maternal blood: Evaluation for trisomy 21 and trisomy 18. *American Journal of Obstetrics and Gynecology, 206*(4), 319. https://doi.org/10.1016/j.ajog.2012.01.030

Takahashi, M., Linh, L. K., Sayed, A. M., Imoto, A., Sato, M., Dila, K. A. S., Huy, N. T., & Moji, K. (2022). Non-invasive prenatal testing (NIPT) implementation in Japan: A comparison with the United Kingdom, Germany, Italy, Sweden, and Taiwan. *International Journal of Environmental Research and Public Health, 19*(24). https://doi.org/10.3390/ijerph192416404

Taylor-Phillips, S., Freeman, K., Geppert, J., Agbebiyi, A., Uthman, O. A., Madan, J., Clarke, A., Quenby, S., & Clarke, A. (2016). Accuracy of non-invasive prenatal testing using cell-free DNA for detection of Down, Edwards and Patau syndromes: A systematic review and meta-analysis. *BMJ Open, 6*(1), Article e010002. https://doi.org/10.1136/bmjopen-2015-010002

Verma, I. C. (2014). Noninvasive prenatal testing: The Indian perspective. *Journal of Fetal Medicine, 1*, 113–118. https://doi.org/10.1007/s40556-014-0025-8

Verma, I. C. (2023). Noninvasive prenatal testing: The Indian perspective. *Journal of Fetal Medicine, 01*(03), 113–118. https://doi.org/10.1007/s40556-014-0025-8

Wright, C. F., & Burton, H. (2009). The use of cell-free fetal nucleic acids in maternal blood for non-invasive prenatal diagnosis. *Human Reproduction Update, 15*(1), 139–151. https://doi.org/10.1093/humupd/dmn047

Yang, J., Chen, M., Shen, W., Wu, H., Shou, J., Sun, J., & Wu, W. (2022). Knowledge, attitudes, and practices of healthcare professionals working in prenatal diagnosis toward expanded non-invasive prenatal testing in China. *Prenatal Diagnosis, 42*(1), 3–14. https://doi.org/10.1002/pd.6075

Zerres, K., Rudnik-Schöneborn, S., & Holzgreve, W. (2021). Do non-invasive prenatal tests promote discrimination against people with Down syndrome? What should be done? *Journal of Perinatal Medicine, 49*(8), 965–971. https://doi.org/10.1515/jpm-2021-0204

Further reading

Xin, X., Wu, S., Xu, H., Ma, Y., Bao, N., Gao, M., Han, X., Gao, S., Zhang, S., Zhao, X., Qi, J., Zhang, X., & Tan, J. (2024). Non-invasive prediction of human embryonic ploidy using artificial intelligence: A systematic review and meta-analysis. *eClinicalMedicine, 77*, Article 102897. https://doi.org/10.1016/j.eclinm.2024.102897

CHAPTER

Advanced genomic methods in non-invasive prenatal screening

3

Charu Rajpal

Department of Biotechnology, Manav Rachna International Institute of Research and Studies, Faridabad, Haryana, India

1. Introduction

The Next-Generation Sequencing (NGS) technology has revolutionized prenatal screening to a new level (Babay et al., 2015). Non-invasive prenatal testing (NIPT) is a highly sensitive and precise method for detecting fetal aneuploidy that relies on cell-free DNA (cfDNA) (Parsaei et al., 2024). Since its inception, NIPT has dramatically enhanced clinical acceptability by demonstrating greater effectiveness than traditional prenatal serum screening methods. Fetal aneuploidy and other chromosomal deformities occurring due to multifactorial reasons affect nine out of 1000 live births (Noia et al., 2026; Albu et al., 2025). Invasive prenatal diagnostic methods like chorionic villus sampling or amniocentesis are applied when biochemical and/or sonographic screenings reveal that a woman has a higher probability of having such aneuploidies. Conventional cytogenetic analysis is used as the gold standard for detecting chromosome abnormalities in fetuses, wherein fetal nucleated cells are cultured via amniocentesis or chorionic villus biopsy. However, the probability of intrusive testing leading to miscarriage is 1% chance. The discovery of free fetal DNA in maternal plasma in 1997 led researchers to develop a reliable method of non-invasive prenatal diagnosis (NIPD) (Ren et al., 2025). According to recent studies, the best way to screen for trisomy 21 is through cell-free DNA in maternal plasma. The detection rate of this approach exceeds 99%, and the false-positive rate exceeds 0.1% (Andonotopo et al., 2025; Lin et al., 2025).

NIPT is one of the most promising areas of molecular medicine research over the last decade. The technology will be cheaper and more efficient, and will even compete with the biochemical screening tests. It will require substantial research to determine whether adding low-cost PAPP-A first-trimester screening (without NT) to NIPT would eliminate the need for a definitive karyotype in both positive and negative outcomes. Non-invasively obtained cell-free fetal DNA (cffDNA) sequencing (fetal genome sequencing or exome sequencing) could only require a few steps to become NIPT technology (Săbău et al., 2025). The capability to routinely analyze a complete fetal genome in a clinical setting will open the door for patients to access information about the fetus far beyond any particular fetal illness, including trisomy. This raises the

Non-Invasive Prenatal Screening of Rare Fetal Genetic Diseases. https://doi.org/10.1016/B978-0-443-27660-6.00006-7

issue that individuals will be exposed to a great deal of information, much of which may be obscure and of minimal therapeutic importance.

When the presence of embryonic disorders has been detected at this young age, the most minimal invasive prenatal therapy (like editing genes through the emerging technology) must be viewed as the second course of action. It is suggested that NIPT should be administered alongside counseling, education, and informed consent. The patients should be provided with genetic counseling and the option of routine confirmatory diagnostic testing when they have aberrant results or other features that indicate a chromosomal issue (Ismayilova et al., 2025). New biomarkers discovered through proteomic advances could also provide a cheaper screening method for NIPT than NGS-based methods, though sensitivity will be reduced. Nevertheless, only new biomarkers can be utilized in the screening procedure, whereas NGS can detect fetal DNA directly, providing actual NIPD that ultimately supersede current invasive biopsy-based diagnostic approaches. This can be done in combination with the size ratio of the cffDNA and the epigenetic methods (Sirica et al., 2025).

NIPT detection and analysis are based on cfDNA isolated from the mother's peripheral blood. New uses and greater effectiveness of NIPT have been achieved in recent years through the purification technologies. The clinical use of NGS has increased the sensitivity of NIPT for detecting aneuploidy. Moreover, it has enabled the identification and analysis of the fetal genome in maternal plasma. Fetal whole exome sequencing can be used to identify the cause of fetal abnormalities with a normal karyotype and to exclude non-invasive diagnoses of hereditary monogenic diseases during pregnancy (Resta et al., 2025). NGS can be a safer method for fetal genetic diagnosis, an essential element of genomic technology development. Besides enhancing screening for fetal abnormalities and the diagnosis of monogenic disorders as the etiology of fetal abnormalities, these new techniques enable NIPD of single-gene disorders with utmost certainty at an early gestational age, without costly testing. This has led to new challenges and new opportunities for scientists and clinicians. Whole-genome, exome, and targeted NGS have not only transformed prenatal genetic testing but also generated large volumes of data previously unheard of, necessitating complex processing and analysis. With these technologies currently being developed in clinical genomics, personalized medicine, and modern healthcare, the field is being revolutionized. Fig. 3.1 depicts the clinical utility workflow of NIPT.

2. Genomic methods in non-invasive prenatal screening

The genomic methods used by NIPS offer significant advantages in terms of accuracy, early detection, and tailored therapy. Careful consideration of these technologies' drawbacks, moral dilemmas, and implementation challenges is required to guarantee their safe and effective application in prenatal screening. To detect fetal aneuploidies and genetic abnormalities, NIPS uses genomic tools to analyze cfDNA in maternal plasma. Prenatal screening for common fetal aneuploidies, such as

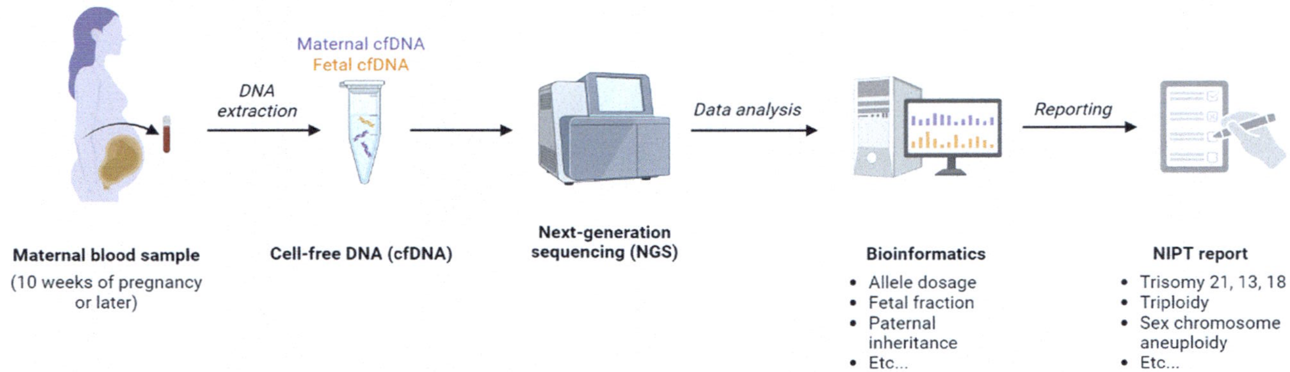

FIGURE 3.1

The clinical workflow of the non-invasive prenatal screening.

trisomy, has become much more accurate and sensitive because of NGS and other genomic approaches. NIPS reduces the risk of miscarriage and does away with the necessity for invasive procedures due to its exceptional detection effectiveness for common chromosomal aneuploidies. Early interventions and customized treatment are made possible by increased prenatal and neonatal screening accuracy (Audibert et al., 2024). It is important to consider both technical and ethical limitations. NIPS, for instance, is extremely sensitive and specific, but it can also have more unknowable and ambiguous effects, which could lead to moral and psychological problems. The high cost and complex sequencing procedures of the existing sequencing techniques limit the availability of NIPS for expectant mothers (Sharma et al., 2024). For high-risk pregnant women, less intrusive procedures are required, which reduces the chance of miscarriage. It is important to take into account both the potential impact of direct marketing to users and the reproductive autonomy of pregnant mothers on the broad adoption of NIPS. In addition to ethical issues, quality control procedures, and the need for protocol standardization, NIPS struggles to evaluate genomic data with ambiguous therapeutic implications.

2.1 Key genomic methods

The main genomic techniques used in NIPS include SNPs, WGS, microarrays, and high-throughput gene sequencing. To analyze sub-chromosomal aneuploidies, these techniques have become more accurate, dependable, and comprehensive. More clinical validation research and comprehensive counseling are required to fully describe this technology, given its ethical concerns and limitations, especially regarding positive predictive values and the potential for misleading outcomes. Advanced genomic methods include high-throughput gene sequencing, WGS, SNPs, microarray, and rolling circle amplification (Iordanescu et al., 2026; Săbău et al., 2025).

Advanced techniques, such as the creation of nanobiosensors and the use of certain nanoparticles, have improved NIPS experiments. As a result, screening tests are now more precise, sensitive, and specific. Emerging genomic technologies, primarily based on NGS, have enabled the study of subchromosomal aneuploidies. This offers a greater diagnosis yield without the miscarriage risks associated with earlier diagnostic techniques (Botos et al., 2023). Although the clinical sensitivity, specificity, and negative predictive value of NIPT approaches are comparable for common fetal trisomies, tests that use SNPs and RCA have demonstrated lower PPV, highlighting the importance of invasive testing to validate positive NIPT results (Iordanescu et al., 2026). Table 3.1 presents genomic methods used in NIPS.

2.2 Role of next-generation sequencing in non-invasive prenatal screening

NGS plays a vital role in non-invasive prenatal screening because it provides a highly sensitive and accurate approach to detecting a spectrum of genetic defects in the fetus. As technology advances, NGS will be a significant contributor to prenatal

Table 3.1 Genomic methods used in non-invasive prenatal screening.

Method	Description	Applications
cffDNA analysis	Analyzing fetal DNA in maternal plasma	Detecting aneuploidies (T21, T18, T13) (Hermawan et al., 2026)
High-throughput sequencing	Next-generation sequencing of cffDNA	High sensitivity and specificity for genetic conditions (Fu & Hu, 2026)
PCR and methylation	Amplification and methylation-specific techniques	Enhancing detection accuracy (Chen et al., 2025)
Size selection	Enriching fetal cfDNA by size	Improving test accuracy (Meng et al., 2025)
Magnetic extraction	Automated extraction of cffDNA	Higher yield of cffDNA (Sheybanikashani et al., 2026)
SPRI	Nanoparticle-enhanced detection of Y-chromosome sequences	Non-invasive fetal sex determination (Calcagno et al., 2021)
Proteomics and transcriptomics	Analyzing proteins and RNA from the placenta	Identifying diagnostic markers (Ge et al., 2023)
Quality control	Use of control materials	Ensuring test reliability (Fu & Hu, 2026)
Genetic counseling	Pre- and post-test counseling	Understanding test results (Diderich et al., 2023)

care, providing expectant parents with more accurate and safer screening tests. NGS reduces the risks to the mother and the fetus by identifying the genetic abnormalities at an early stage without involving intrusive therapies. It is more efficient, highly sensitive, and specific than traditional prenatal serum screening techniques for identifying fetal aneuploidy. The approach could reveal more genetic issues, including rare conditions and structural variants. Some of the challenges include process standardization, quality assurance, and ethical issues. NGS is not widely used due to its high cost, especially in resource-limited settings. To deploy NIPT technologies as diagnostic screening tests for all pregnancies in the near future, their accuracy and cost must be improved (Ding et al., 2026). NGS has greatly enhanced the detection rate of common trisomies and sex chromosomal aneuploidies compared with traditional screening techniques. It offers a better diagnostic yield with no miscarriage problems that were associated with the earlier ratified diagnostic assays (Findley & Northrup, 2021).

3. Single-nucleotide polymorphism analysis in non-invasive prenatal screening

Single-nucleotide polymorphism (SNP) analysis has become a crucial technique in NIPS for identifying a wide range of genetic diseases. This method detects possible aneuploidies and other genetic conditions by comparing SNPs from the mother and the fetus. Fig. 3.2 provides a representative depiction of SNP analysis.

SNP-based NIPS accurately detects trisomies 13, 18 and 21. For example, SNP-based techniques were found to be 100% accurate in all cases of trisomy 13 in euploid samples and in samples that met the quality criteria. In another study, SNP-based NIPS was discovered to have a high positive prediction value (PPV) of trisomy 18 (100%) and trisomy 21 (94%).

This method also improves method output using high-depth resequencing and can identify microdeletion diseases, including 22q11.2 deletion syndrome. This approach influences positive predictive value. NIPS from SNPs has been observed in clinical research to detect microdeletions (22q11.2, Prader–Willi, Angelman, and cri-du-chat syndromes) with high detection rates and relatively low false-positive rates (Cassou, 2025). However, this method is not effective when the fetal fraction is low. Research indicates that whole-genome sequencing (WGS) can achieve reasonable specificity and sensitivity even at low fetal fractions. Nevertheless, the SNP techniques would cease to be useful at a fetal composition of less than approximately 3% (Grendár et al., 2019). NIPS systems based on SNP technology are also suitable for prenatal testing, having low false-positive rates and high sensitivity rates in most genetic diseases.

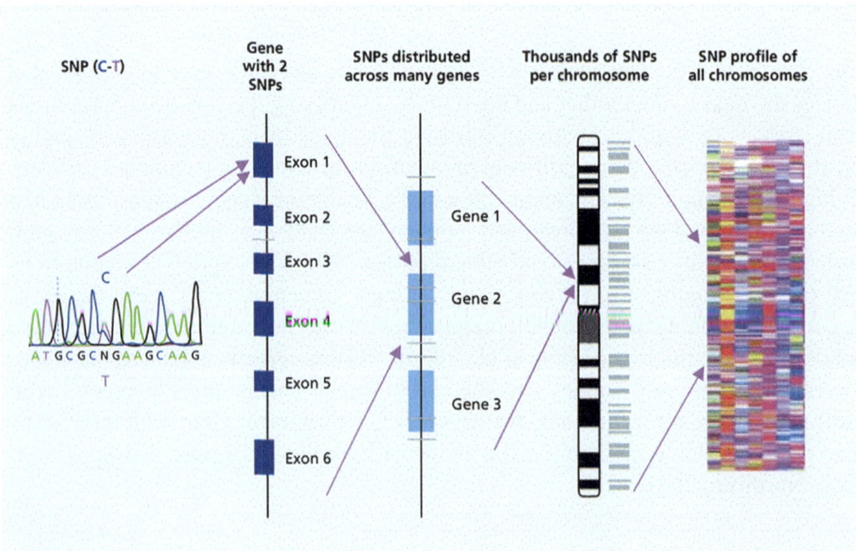

FIGURE 3.2

A representative diagram showing single nucleotide polymorphism.

4. Comparative Genomic Hybridization (aCGH) in non-invasive prenatal screening

aCGH Array (Comparative Genomic Hybridization) is a high-resolution molecular cytogenetic method that has had an important impact on prenatal genetics and has recently been mentioned in the context of the development of advanced NIPS. In the past, aCGH has been applied to invasive prenatal diagnostic techniques like amniocentesis and chorionic villus sampling, where fetal DNA is simply collected and subjected to analysis of chromosomal abnormalities. As non-invasive genomic technologies that examine circulating cfDNA in maternal plasma were developed, however, investigators have considered the conceptual and experimental redesign of aCGH for non-invasive prenatal use (Choi et al., 2026).

The basic idea of aCGH is the relative hybridization of test DNA labeled with fluorescent dyes and reference DNA onto microarray systems with thousands to millions of genomic probes spanning the entire genome. Fluorescence changes in the ratio of specific intensities can be used to detect genomic gains and losses, as well as copy number variations (CNVs), microduplications, and microdeletions, with far greater sensitivity than standard karyotyping. The low levels of fetal fraction (usually between 3 and 15% of total circulating DNA in maternal blood) and the fragmented nature of cffDNA are a challenge to the integration of aCGH in NIPS. This low percentage of fetal DNA requires very sensitive laboratory techniques, including whole-genome amplification, optimized probe hybridization, and highly specialized normalization algorithms to amplify the fetal genomic signal above the maternal background noise. The resolution and analytical sensitivity of aCGH have been enhanced significantly with the creation of high-density oligonucleotide arrays by biotechnology companies like Agilent Technologies and developers of sequencing platforms, like Illumina, which enable smaller chromosomal adaptations of comparable clinical significance during prenatal development. aCGH provides an analysis of copy numbers on a more detailed scale, and it is especially useful when it comes to larger panels of prenatal screening markers to identify rare genomic diseases. Experimental non-invasive methods, such as cffDNA in maternal plasma, may be amplified and hybridized to microarrays, and the hybridization intensity may be computed to detect dosage imbalances between genomic regions. It has been extensively dependent on bioinformatics tools and enhanced software that perform normalization, segmentation, and aberration detection to differentiate true fetal CNVs from maternal genomic variation and technical artifacts. Companies like Agilent CytoGenomics and BlueFuse Multi offer proprietary analysis systems that are typically applied to process array data, produce copy number plots, and visualize prenatal sample chromosomal gains and losses (Liu et al., 2025).

Also, genome annotation tools, such as the UCSC Genome Browser, and clinical repositories, such as ClinVar and DECIPHER, are vital for interpreting identified CNVs and evaluating their clinical importance for prenatal reporting and genetic counseling. This is because aCGH is one of the greatest advantages of the technique in the prenatal setting for identifying subchromosomal abnormalities that may not substantially alter overall chromosome dosage but can have significant clinical consequences, thereby yielding a higher diagnostic yield than traditional cytogenetic methods. Nevertheless, the technology remains technically challenging for its application in NIPS, as it is characterized by maternal DNA contamination, hybridization noise, amplification bias, and relatively higher DNA input requirements compared to sequencing-based NIPT technologies. The short fragmentation of cffDNA, about 140–200 base pairs, can also influence the hybridization efficiency of microarray probes, requiring probe density optimisation and sophisticated signal correction algorithms to ensure reliable CNV detection. Moreover, aCGH is unable to identify balanced chromosomal rearrangements, single-nucleotide variants, or very low levels of mosaicism, which limits its diagnostic abilities alone and commonly necessitates secondary genomic methods to analyze SNPs or NGS (Liu et al., 2025). In more sophisticated study systems, hybrid schemes combining shallow WGS and aCGH have been proposed, with sequencing used for determining fetal fraction and initial screening for aneuploidy, and aCGH for high-resolution validation and characterization of CNVs. The diagnostic power of chromosomal microarray analysis in invasive prenatal testing has been recognised by professional organisations, including the International Society of Prenatal Diagnosis, which indirectly recommends the use of comparable high-resolution genomic technologies in future non-invasive screening paradigms. Moreover, the algorithms used in segmentation, like the circular binary segmentation and hidden Markov models, are used in the analysis of the aCGH data to precisely detect chromosomal imbalances based on the data on the intensity of the hybridization, particularly in the context of low-input DNA samples, as is common in the maternal plasma. An innovation, which is cell-based NIPT (cbNIPT), utilizes intact fetal trophoblast cells obtained in maternal blood are analyzed by aCGH, which can effectively overcome the low-fetal-fraction limitation and is able to detect high-resolution genome-wide CNVs in a non-invasive fashion (Gyokova et al., 2026).

The ethical aspect should also be taken into consideration when applying high-resolution genomic tools, such as aCGH, in prenatal screening, because the greater the detection capability, the greater the likelihood of identifying variants of uncertain significance, which require close interpretation and genetic counseling to avoid undue anxiety in parents (Fiorentino, 2023). The next generation of microarray sensitivity, fetal DNA enrichment methods, artificial intelligence-based signal processing, and combined bioinformatics workflows could further enhance the feasibility of including aCGH in non-invasive prenatal screening protocols (Arumugam et al., 2024).

5. Major differences between aCGH and other non-invasive methods of prenatal screening

aCGH is more extensive than other chromosomal analyses due to its ability to determine sub-microscopic copy number variations throughout the whole genome in a single experiment. aCGH is a high-resolution method that can detect submicroscopic microdeletions or microduplications compared to older methods. It can also identify numerical and structural imbalances in chromosomes that are less than 100 kilobases. The advantages of ACG over other non-invasive methods of prenatal screening include: Quick Results: aCGH results can be obtained within 3 days, compared to normal G-banded analysis, which takes a minimum of 2 weeks to complete. Better diagnosis: Compared with conventional karyotyping, aCGH has proved more helpful for diagnosing harmful chromosomal abnormalities in the prenatal context, offering greater genetic data, particularly in fetuses with structural sonographic appearances. Comprehensive Genome-Wide Screening: Comprehensive genome-wide screening of segmental genomic copy number alterations is increasingly becoming an indispensable clinical diagnostic method, slowly replacing cytogenetic methods (Liu et al., 2025). The main differences, benefits, and drawbacks of aCGH compared to other non-invasive prenatal screening techniques are presented in Table 3.2.

6. Droplet digital PCR

Droplet digital PCR (ddPCR) is a complete molecular technology that can be used to identify nucleic acids. The high sensitivity and the ability to report absolute values

Table 3.2 Key differences between aCGH and other non-invasive prenatal screening methods.

Feature	aCGH	NIPS
Resolution	High-resolution, detects submicroscopic CNVs	Lower resolution, primarily detects common aneuploidies (Zhang et al., 2020)
Detection capability	Detects unbalanced structural and numerical abnormalities	Detects common aneuploidies, limited to subchromosomal defects (Lu et al., 2023)
Invasiveness	Invasive (requires CVS or amniocentesis)	Non-invasive (analyzes cfDNA in maternal blood) (Arumugam et al., 2024)
Turnaround time	Rapid (results within 3 days)	Quick, but depends on the lab and test specifics (Zhang et al., 2020)
Clinical utility	Useful for detailed genomic analysis in abnormal fetuses	Effective for screening common aneuploidies, reducing invasive tests (Zhang et al., 2020)
Limitations	Cannot detect balanced rearrangements, some ploidies	Limited detection beyond common aneuploidies, lower resolution (Lu et al., 2023)

for low-abundance targets in DdPCR render it a promising method for NIPT. Fetal aneuploidies such as Trisomy 21 have been sensitively and accurately detected. The method has also been used to accurately identify fetal paternity mutations in NIPD of single-gene diseases (Guo et al., 2025). The ddPCR is a dependable, non-invasive prenatal diagnostic tool because it offers the best sensitivity, accuracy, and consistency (Dai et al., 2024). Nevertheless, cffDNA multiplex detection via ddPCR can lack adequate specificity due to complex primer design. Ambiguous results are highly probable, particularly when rare Mendelian diseases are under study, which may make therapy implementation difficult (Shaw et al., 2023). ddPCR is an excellent alternative to prenatal diagnostics, as it has been shown to be equally precise as NGS at a lower cost and complexity. The validation of the methodology used to identify maternally and paternally inherited fetal alleles in maternal plasma has been a step in the use of NIPD in all pregnancies, regardless of the disease's parental lineage (Zhou & Jiang, 2023). Using ddPCR as an NIPD method has several barriers, including the need to carefully scrutinize the opposing data and reduce the rate of inconclusive results. DDPCR should be improved in terms of detection sensitivity, optimisation of multi-target analysis, cost-saving, simplification of operations, and standardization to increase the clinical utility of the technique in NIPT (Zhang et al., 2025).

7. Bioinformatics and data analysis

GenomeMixer and TRUST are bioinformatic tools designed to predict fetal fraction and sequencing coverage, thereby enhancing the accuracy of NIPT data. The methods analyze computational NIPT (including RAPIDR, NIPTeR, and WisecondorX) owing to sensitivity to sequencing depth (Teder, 2022). To interpret NIPT data and ensure its reliability, calculations of fetal proportion and sequencing depth are necessary. In a bid to make NIPT data results easier to comprehend and visualize, a web-based application, NIPTviewer, was developed to facilitate clinical interpretation of test results (Smeds et al., 2025). Two technological limitations that inhibit NIPT are the importance of sequencing depth for trisomy detection and the need for uniform laboratory methods. It is the responsibility of monitoring and verifying the accuracy of NIPT since low fetal fraction and maternal genetic disorders may result in false-negative findings (Xiang et al., 2025). Lastly, NIPT employs bioinformatics techniques, particularly NGS, to detect genetic defects and fetal aneuploidies. NIPT results can be evaluated and verified using applications and data analysis tools such as NIPTviewer. The necessity of homogeneity and the technical constraints, however, are major challenges. Ethical issues surrounding prenatal and neonatal screening focus on the proper interpretation of data and the broader implications for genetic information.

8. Clinical validation and implementation

The pillars of the ethical framework for prenatal screening are, the purpose of screening, the proportionality of the test, justice, and societal considerations

(Morales de Machín & Machín Cáceres, 2024). Some of the challenges include concerns about informed choices, the pre-test counseling process, financial issues, and potential discrimination against people with disabilities. It is implied that NIPT must only generate test results relevant to reproductive decision-making, that adequate pre-test counseling must be provided so that informed choice can be made, and that equal access must be secured (Kater-Kuipers et al., 2018; Marton et al., 2025).

9. Clinical and ethical considerations

Cost and Accessibility: As more people use NIPT, its high cost should go down, making it more affordable.

Quality Control: Strict testing procedures must be followed to ensure the precision and reliability of NIPT results.

Ethical Concerns: Genetic counseling must carefully address the psychological effects on patients as well as the possibility of unclear outcomes (Hayward & Chitty, 2018).

10. Future directions

To increase detection rates and reduce unnecessary testing, NIPT is being integrated into national screening programs, such as the NHS in the UK. As genome sequencing and bioinformatics continue to advance, NIPT's precision and reach will increase, solidifying its position as a vital component of prenatal care (Moustakli et al., 2024). Hence, emerging technologies in non-invasive prenatal screening, particularly those based on NGS, are revolutionizing prenatal care by providing safer, more comprehensive, and earlier detection of genetic abnormalities. However, these advancements come with challenges that require careful consideration of clinical, ethical, and regulatory aspects to maximize their benefits (Maternal et al., 2024).

References

Albu, C. C., Brăila, A. D., Poalelungi, C. V., Bohîltea, L. C., Bănățeanu, A. M., Damian, C. M., & Bogdan-Andreescu, C. F. (2025). Micrognathia as a diagnosis marker for the prenatal identification of edwards syndrome. *Biomedicines, 13*(3), 573.

Andonotopo, W., Bachnas, M. A., Pribadi, A., Alamsyah Azis, M., Aldika Akbar, M. I., Ernawati, & Kurjak, A. (2025). Integrating NIPT and ultrasound for detecting foetal aneuploidies and abnormalities. *Journal of Perinatal Medicine, 53*(7), 789–802.

Arumugam, S., Kalluri, S. S., Sharmila, V., Subbiah, N. K., Mocherla, A., Kulkarni, J., & Ghoshal, J. A. (2024). Acceptability of prenatal screening tests among expectant mothers in India: Insights and implications for public health. *Cureus, 16*(5).

Audibert, F., Wou, K., Okun, N., De Bie, I., & Wilson, R. D. (2024). Guideline No. 456: Prenatal screening for foetal chromosomal anomalies. *Journal of Obstetrics and Gynaecology Canada, 46*(11), Article 102694.

Babay, L. E., Horanyi, D., Rigo Jr, J., & Nagy, G. R. (2015). Next generation sequencing and its applications in NIPT of aneuploidies. *Orvosi Hetilap, 156*(26), 1041–1048.

Botos, L., Szatmári, E., & Nagy, G. R. (2023). Prenatal and postnatal genetic testing toward personalized care: The non-invasive perinatal testing. *Molecular and Cellular Probes, 72*, Article 101942.

Calcagno, M., D'Agata, R., Breveglieri, G., Borgatti, M., Bellassai, N., Gambari, R., & Spoto, G. (2021). Nanoparticle-enhanced surface plasmon resonance imaging enables the ultrasensitive detection of non-amplified cell-free foetal DNA for NIPT. *Analytical chemistry, 94*(2), 1118–1125.

Cassou, M. (2025). Dépistage prénatal des trisomies: qu'en pensent les personnes concernées? *Droit, Santé et Société, 12*(1–2), 47–51.

Chen, J., Wang, Z., & Yang, B. (2025). mpMRI-based MGMT methylation status prediction for glioblastoma through off-the-shelf deep features: A multi-dataset feasibility study. *Journal of Applied Clinical Medical Physics, 26*(12), Article e70373.

Choi, N., Kim, H. Y., & Ko, J. M. (2026). Clinical and genetic significance of chromosomal microarray screening of asymptomatic newborns. *Journal of Clinical Laboratory Analysis*, Article e70160.

Dai, P., Chen, C., Zhao, G., Liu, N., & Kong, X. (2024). Expert consensus on the application of digital PCR non-invasive prenatal screening technology for the preliminary implementation of foetal chromosomal disease screening. *Zhonghua yi xue yi chuan xue za zhi= Zhonghua yixue yichuanxue zazhi= Chinese journal of medical genetics, 41*(10), 1164–1170.

Diderich, K. E., Klapwijk, J. E., van der Schoot, V., Brüggenwirth, H. T., Joosten, M., & Srebniak, M. I. (2023). Challenges and pragmatic solutions in pre-test and post-test genetic counseling for prenatal exome sequencing. *The Application of Clinical Genetics*, 89–97.

Ding, Y., Ni, K., Fan, X., & Yan, Q. (2026). A Bayesian decision-theoretic optimization model for personalized timing of NIPT based on maternal BMI. *Mathematics, 14*(3), 437.

Findley, T. O., & Northrup, H. (2021). The current state of prenatal detection of genetic conditions in congenital heart defects. *Translational Pediatrics, 10*(8), 2157.

Fiorentino, D. (2023). Prenatal screening for microdeletions and rare autosomal aneuploidies. *Clinical Obstetrics and Gynecology, 66*(3), 579–594.

Fu, G., & Hu, S. (2026). Clinical application of fetal Nuchal Translucency combined with noninvasive prenatal testing in screening chromosome abnormalities. *PLoS One, 21*(3), Article e0344739.

Ge, L., Huang, P., Miao, H., Yu, H., Wu, D., Chen, F., & Hua, J. (2023). The new landscape of differentially expression proteins in placenta tissues of gestational diabetes based on iTRAQ proteomics. *Placenta, 131*, 36–48.

Grendár, M., Loderer, D., Laučeková, Z., Švecová, I., Hrtánková, M., Horňáková, A., & Danko, J. (2019). Uncertainty of foetal fraction determination in Non-Invasive Prenatal screening by highly polymorphic SNPs. *Journal of Biotechnology, 299*, 32–36.

Guo, Y., Charoenkwan, P., Traisrisilp, K., Piyamongkol, W., & Tongprasert, F. (2025). Application of digital polymerase chain reaction (dPCR) in NIPT (NIPT). *Biomolecules, 15*(3), 360.

Gyokova, E., Hristova-Atanasova, E., Odumosu, E., & Dimitrova, K. (2026). Advancing prenatal diagnosis: From conventional karyotyping to genome-wide CNV analysis. *Life, 16*(2), 309.

Hayward, J., & Chitty, L. S. (April 2018). Beyond screening for chromosomal abnormalities: Advances in non-invasive diagnosis of single gene disorders and foetal exome sequencing. In *Seminars in foetal and neonatal medicine, 23* (pp. 94–101). WB Saunders.

Hermawan, B. J., Utama, B. I., & Putra, A. E. (2026). Identification of gender in early pregnancy using cell-free foetal Dna. *Andalas Obstetrics And Gynecology Journal, 10*(1), 135–144.

Iordanescu, I. I., Catana, A., Barabas Cuzmici, Z., Chelu, P., Basangiu, B. F., Severin, E., & Militaru, M. S. (2026). Prenatal diagnosis of sex chromosome aneuploidies: A retrospective study using QF-PCR, SNP-Based chromosomal microarray analysis, and NIPT. *Genes, 17*(2), 171.

Ismayilova, M., Hasanova, A., & Semikhodskii, A. (2025). Preimplantation testing of human blastomeres for aneuploidy increases IVF success in couples where Male partners had abnormal semen parameters. *Biomedicines, 13*(5), 1191.

Kater-Kuipers, A. B. E. M., Bunnik, E. M., De Beaufort, I. D., & Galjaard, R. J. H. (2018). Limits to the scope of NIPT (NIPT): An analysis of the international ethical framework for prenatal screening and an interview study with Dutch professionals. *BMC Pregnancy and Childbirth, 18*(1), 409.

Lin, K., Xu, J., Su, H., Wang, T., Nie, Q., Fang, Q., & Ren, B. (2025). A multiplex digital PCR-based all-in-one non-invasive prenatal test for simultaneous detection of foetal chromosomal aneuploidies and foetal fraction quantification. *Microchemical Journal*, Article 116569.

Liu, L., She, L., Zheng, Z., Huang, S., & Wu, H. (2025). Chromosomal microarray analysis and karyotype analysis for prenatal diagnosis of fetuses with abnormal ultrasound soft markers. *Journal of Clinical Laboratory Analysis, 39*(10), Article e70033.

Lu, S., Kakongoma, N., Hu, W. S., Zhang, Y. Z., Yang, N. N., Zhang, W., & Zhang, Z. F. (2023). Detection rates of abnormalities in over 10,000 amniotic fluid samples at a single laboratory. *BMC Pregnancy and Childbirth, 23*(1), 102.

Marton, T., Erdélyi, Z. R., Takai, M., Mészáros, B., Supák, D., Ács, N., & Valent, S. (2025). Systematic review of accuracy differences in NIPT methods for common aneuploidy screening. *Journal of Clinical Medicine, 14*(8), 2813.

Maternal, R. H. B. O. C., Xu, C., Yin, C., Yin, A., Liu, S., Jiang, Y., & Child Health Association. (2024). Expert consensus on the clinical application strategy of NIPT2. 0, a new-generation non-invasive prenatal screening technology. *Zhonghua yi xue yi chuan xue za zhi= Zhonghua yixue yichuanxue zazhi= Chinese journal of medical genetics, 41*(10), 1155–1163.

Meng, Y., Yan, Y., Yue, X., Li, Y., & Mo, Y. (2025). Advances in cfDNA research for pregnancy-related diseases. *Frontiers in Cell and Developmental Biology, 13*, Article 1600532.

Morales de Machín, A., & Machín Cáceres, E. (2024). Prenatal screening for aneuploidies by analysis of circulating total free deoxyribonucleic acid in maternal plasma. Narrative review. *Revista de Obstetricia y Ginecologia de Venezuela, 84*(2), 185–204.

Moustakli, E., Zikopoulos, A., Skentou, C., Bouba, I., Dafopoulos, K., & Georgiou, I. (2024). Evolution of minimally invasive and non-invasive preimplantation genetic testing: An overview. *Journal of Clinical Medicine, 13*(8), 2160.

Noia, G., Pasciuto, T., Ria, F., Pontecorvi, A., Sacco, M., Teveroni, E., & De Santis, M. (2026). A new hypothesis on the etiology of Down syndrome: The role of anti-zona pellucida antibodies as an age-independent factor. *International Journal of Molecular Sciences, 27*(2), 991.

Parsaei, M., Dashtkoohi, M., Salmani, T. A., Najafi, M. S., Haddadi, M., Ghaemi, M., & Hantoushzadeh, S. (2024). Potential efficacy of digital polymerase chain reaction for non-invasive prenatal screening of autosomal aneuploidies: A systematic review and meta-analysis. *BMC Pregnancy and Childbirth, 24*(1), 472.

Ren, J., Guan, X., Lv, W., Yan, Y., Si, Y., Yang, S., & Yin, C. (2025). A retrospective analysis of 38,652 amniotic fluid karyotype. *Frontiers in Genetics, 16*, Article 1655290.

Resta, C., Xiong, R., Sturrock, S., Story, L., Ntalianis, K. M., Seed, P., & Oteng-Ntim, E. (2025). NIPT for the diagnosis of sickle cell disease in high-risk pregnancies: A systematic review and statistical summary of the current literature. *European Journal of Obstetrics & Gynecology and Reproductive Biology*, Article 114799.

Săbău, I. D., Bohîlțea, L. C., Țurcan, M., Gheorghe, A. S., Riza, M., Mitroi, M., & Ceausu, I. (2025). Comprehensive prenatal genetic analysis: From NIPT to whole-exome sequencing in a high-risk pregnancy with gaucher disease—A case report and literature review. *Journal of Mind and Medical Sciences, 12*(1), 25.

Sharma, I., Negi, N., & Saha, S. C. (2024). Sequencing: A promising path in the detection of foetal health. In *Non-invasive prenatal screening (NIPS) in clinical practice* (pp. 273–288). Singapore: Springer Nature Singapore.

Shaw, J., Scotchman, E., Paternoster, B., Ramos, M., Nesbitt, S., Sheppard, S., & Chandler, N. (2023). Non-invasive foetal genotyping for maternal alleles with droplet digital PCR: A comparative study of analytical approaches. *Prenatal Diagnosis, 43*(4), 477–488.

Sheybanikashani, S., Zhou, J., & Papautsky, I. (2026). Blood microfluidics: Progress and challenges. *Lab on a Chip, 26*, 1123–1147. https://doi.org/10.1039/D5LC00059A

Sirica, R., Ottaiano, A., D'Amore, L., Ianniello, M., Petrillo, N., Ruggiero, R., & Fico, A. (2025). Advancing non-invasive prenatal screening: A targeted 1069-gene panel for comprehensive detection of monogenic disorders and copy number variations. *Genes, 16*(4), 427.

Smeds, P., Baranowska Körberg, I., Melin, M., & Ladenvall, C. (2025). Visualization using NIPTviewer support the clinical interpretation of noninvasive prenatal testing results. *BMC Medical Genomics, 18*(1), 15.

Teder, H. (2022). *Developing computational methods and workflows for targeted and whole-genome sequencing-based NIPT.*

Xiang, J., Sun, X., Peng, J., Zhang, H., Shen, J., Li, J., & Peng, Z. (2025). Determining the origin of genome aberrations improves the positive predictive value of NIPT for 22q11. 2 deletion syndrome. *Scientific Reports, 15*(1), Article 24755.

Zhang, X., Wang, Y., Zhao, N., Liu, P., & Huang, J. (2020). Variations in chromosomal aneuploidy rates in IVF blastocysts and early spontaneous abortion chorionic villi. *Journal of Assisted Reproduction and Genetics, 37*(3), 527–537.

Zhang, Z., Liu, X., Peng, C., Du, R., Hong, X., Xu, J., & Liu, D. (2025). Machine learning-aided identification of faecal extracellular vesicle microRNA signatures for noninvasive detection of colorectal cancer. *ACS Nano, 19*(10), 10013–10025.

Zhou, Y., & Jiang, Y. (2023). Current advances in genetic testing for spinal muscular atrophy. *Current Genomics, 24*(5), 273–286.

CHAPTER

4

Trisomy disorders: trisomy 21 (Down's syndrome), Edwards syndrome (trisomy 18), and Patau syndrome (trisomy 13)

Bekele Chakiso[1], **Riyaz Ahmad Rather**[2]

[1]*Department of Biology, College of Natural and Computational Science, Wachemo University, Hossana, Ethiopia;* [2]*Department of Bioinformatics, Faculty of Engineering & Technology, Marwadi University, Rajkot, Gujarat, India*

1. Introduction

Trisomy disorders constitute a major category of chromosomal aneuploidies. They result from the presence of an extra copy of an autosome, which leads to a significant disturbance in embryonic development, organogenesis, and postnatal survival (Hassold & Hunt, 2001). Among these, Trisomy 21 (Down syndrome), Trisomy 18 (Edwards syndrome), and Trisomy 13 (Patau syndrome) represent the most frequently encountered and clinically significant autosomal trisomies compatible with live birth (Wu et al., 2013). These conditions contribute to global perinatal morbidity, neonatal mortality, long-term disability, and psychosocial burden for affected families and healthcare systems (Springett & Morris, 2014).

Trisomy disorders occur when chromosomal segregation during meiosis occur wrongly, most commonly nondisjunction events occurring in maternal gametogenesis. The resulting gene dosage imbalance disrupts tightly regulated developmental pathways. The altered pathway produces complex and multisystem phenotypes that vary in severity depending on the specific chromosome involved, the extent of mosaicism, and associated genetic or environmental modifiers (Antonarakis et al., 2020). Advances in cytogenetics and molecular biology over the past 5 decades have considerably refined our understanding of these disorders. These syndromes are now understood as well-characterized genomic conditions with identifiable molecular mechanisms.

Epidemiologically, Trisomy 21 is the most prevalent, with an estimated global incidence of approximately one in 700–800 live births, whereas Trisomy 18 and Trisomy 13 are comparatively rarer (Morris et al., 2008). They occur in approximately one in 5000 and one in 10,000–16,000 live births, respectively (Srinivasan et al., 2025). Despite their lower incidence, Edwards and Patau syndromes are associated

Non-Invasive Prenatal Screening of Rare Fetal Genetic Diseases. https://doi.org/10.1016/B978-0-443-27660-6.00009-2

with disproportionately high rates of fetal loss, stillbirth, and early neonatal death (Carey, 2021, pp. 937–956). Maternal age remains the most consistently identified risk factor across all three conditions, although paternal age, environmental exposures, and stochastic meiotic errors also contribute to risk (Sherman et al., 2007).

The clinical manifestations of trisomy have a broad phenotypic spectrum, encompassing structural congenital anomalies, neurodevelopmental impairment, growth restriction, and organ-specific dysfunctions. Trisomy 21 is associated with intellectual disability of variable severity, characteristic craniofacial features, congenital heart disease, and increased susceptibility to endocrine and hematological disorders. In contrast, Trisomy 18 and Trisomy 13 present with more severe malformations, including central nervous system anomalies, complex cardiac defects, and craniofacial abnormalities, limiting long-term survival (Rasmussen et al., 2003).

The diagnosis of trisomy disorders is conventionally carried using invasive prenatal procedures such as amniocentesis and chorionic villus sampling, coupled with conventional karyotyping. While diagnostically definitive, these procedures carry procedure-related miscarriage risks (Akolekar et al., 2015). Non-invasive prenatal screening (NIPS) is an emerging technology based on analysis of cell-free fetal DNA used as prenatal risk assessment for common trisomies (van't Oever et al., 2025; De Jong et al., 2010). In this chapter, we provide an overview of Trisomy 21, Trisomy 18, and Trisomy 13, their genetic foundations, clinical features, diagnostic strategies, management approaches, and ethical considerations.

2. Genetic and biological basis of trisomy disorders

Trisomy disorders arise from numerical chromosomal abnormalities in which an individual possesses three copies of a specific autosome instead of the normal diploid complement. This condition is also termed autosomal aneuploidy. It disrupts genomic balance and alters tightly regulated gene expression networks essential for normal embryonic and fetal development. The pathogenic consequences of trisomy are primarily due to gene dosage imbalance, rather than the presence of structurally abnormal genes, making trisomy disorders paradigmatic examples of dosage-sensitive genomic diseases (Birchler & Veitia, 2012). Table 4.1 provides the clinical and genetic comparison of the three major autosomal trisomy syndromes.

2.1 Chromosomal nondisjunction and meiotic errors

The most common genetic mechanism underlying trisomy disorders is chromosomal nondisjunction (Fig. 4.1). It is a failure of homologous chromosomes or sister chromatids to segregate properly during meiosis. Nondisjunction may occur during meiosis I, resulting in gametes containing both homologous chromosomes, or during meiosis II, leading to the duplication of sister chromatids. Fertilization involving such aneuploid gametes produces a zygote with an extra chromosome, which is subsequently propagated to all somatic cells in cases of full trisomy (Webster & Schuh,

Table 4.1 Clinical and genetic comparison of the three major autosomal trisomy syndromes.

Feature	Trisomy 21	Trisomy 18	Trisomy 13
Chromosome involved	Chromosome 21	Chromosome 18	Chromosome 13
Incidence	1 in 700–800	1 in 5000–6000	1 in 10,000–16,000
Genetic mechanism	Mostly full trisomy	Mostly full trisomy	Full trisomy or Robertsonian translocation
Maternal age effect	Strong	Strong	Strong
Growth restriction	Mild to moderate	Severe	Severe
Congenital heart disease	40%–50%	>90%	>80%
Neurological impairment	Mild–moderate intellectual disability	Severe	Profound
Characteristic features	Flat facial profile, hypotonia	Clenched hands, rocker-bottom feet	Holoprosencephaly, cleft lip
Survival	Into adulthood	<10% survive 1 year	<5–10% survive 1 year
Prognosis	Compatible with long life	Poor	Very poor

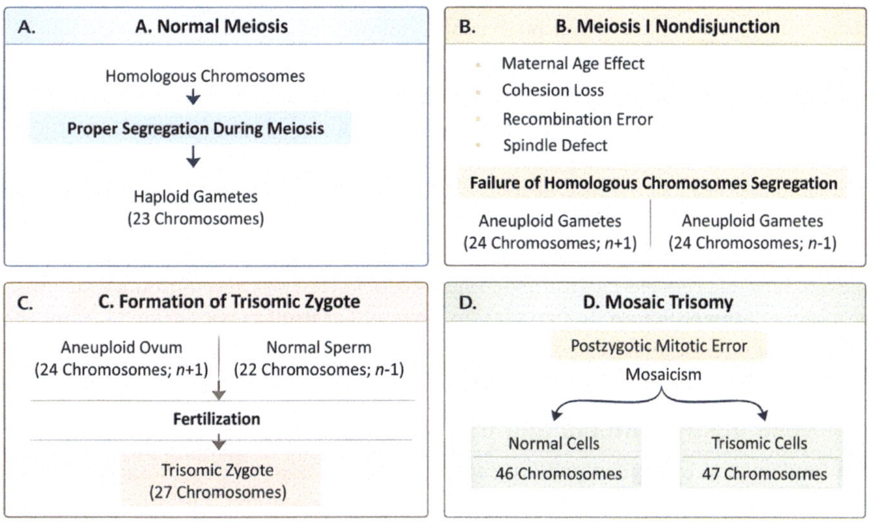

FIGURE 4.1

Mechanisms underlying the formation of autosomal trisomy.

2019). Extensive cytogenetic analyses have demonstrated that the majority of trisomy cases (particularly Trisomy 21) originate from maternal meiotic nondisjunction, with meiosis I errors accounting for a substantial proportion. This maternal predominance is attributed to prolonged oocyte arrest in prophase I, which may last for decades, increasing susceptibility to recombination errors, cohesion loss, and spindle assembly defects (Jones & Lane, 2013). In contrast, paternal nondisjunction events are relatively rare but have been documented, particularly in mosaic or postzygotic cases.

2.2 Role of recombination and chromosome-specific vulnerability

Aberrant meiotic recombination plays a critical role in trisomy formation. Normal recombination provides proper chromosomal pairing and segregation; however, reduced recombination frequency, misplaced crossover events, or absence of chiasmata can destabilize homolog alignment and increase nondisjunction risk. Certain chromosomes, including chromosomes 13, 18, and 21, appear particularly vulnerable to these errors due to their acrocentric structure, smaller size, and specific recombination landscapes (Lamb et al., 2005). Chromosome 21, the smallest human autosome, exhibits a limited recombination domain, making it especially prone to segregation errors. Similarly, chromosomes 13 and 18 contain gene-dense regions essential for early development, which partially explains the severe phenotypic consequences observed when trisomy occurs (Mehler & Mattick, 2007).

2.3 Mosaicism and postzygotic mitotic errors

Not all trisomy disorders involve uniform chromosomal duplication across all cells. Mosaic trisomy results from postzygotic mitotic nondisjunction, producing a mixture of euploid and aneuploid cell lines within the same individual. The clinical severity of mosaic trisomy depends on the proportion, distribution, and tissue-specific localization of trisomic cells (Levy et al., 2021). Mosaicism is more frequently observed in Trisomy 21 and is associated with milder phenotypes compared to complete trisomy, although significant variability exists. Mitotic rescue mechanisms, such as trisomy rescue, may also occur during early embryogenesis, whereby loss of the supernumerary chromosome restores euploidy in some cell lines. While this process can mitigate phenotypic severity, it may introduce secondary complications such as uniparental disomy, which itself carries clinical implications depending on the chromosome involved (Robinson, 2000).

2.4 Gene dosage effects and transcriptomic dysregulation

The fundamental biological consequence of trisomy is altered gene dosage, leading to overexpression of genes located on the affected chromosome. This imbalance disrupts cellular homeostasis and perturbs regulatory networks far beyond the trisomic chromosome itself. Transcriptomic studies have shown that trisomy induces global gene expressionchanges, affecting genes on other chromosomes through secondary regulatory effects, epigenetic modifications, and altered transcription factor availability (Letourneau et al., 2014). In Trisomy 21, for example, overexpression of dosage-sensitive genes involved in neural development, oxidative stress, and cell cycle regulation contributes to characteristic neurodevelopmental and systemic manifestations. Comparable mechanisms operate in Trisomy 18 and Trisomy 13, where dysregulation of developmental signaling pathways leads to profound structural anomalies and impaired organogenesis (Volk et al., 2017).

2.5 Epigenetic and developmental consequences

Beyond gene dosage, trisomy is associated with widespread **epigenetic alterations**, including changes in DNA methylation, histone modifications, and chromatin architecture. These epigenetic disruptions can amplify transcriptional imbalance and interfere with lineage specification during embryonic development (Mendioroz et al., 2015). Experimental models suggest that trisomy affects cell proliferation rates, increases apoptosis, and alters cellular differentiation trajectories, particularly in neural crest cells and cardiac progenitors. Collectively, these genetic and biological mechanisms explain both the shared features and syndrome-specific variability observed across Trisomy 21, 18, and 13. Understanding these foundational processes is essential for interpreting phenotypic heterogeneity, improving prenatal risk stratification, and developing targeted therapeutic or supportive interventions.

3. Trisomy 18

Edwards syndrome is a severe autosomal aneuploidy characterized by the presence of an extra copy of chromosome 18. First described by John Hilton Edwards in 1960, the syndrome is associated with profound developmental abnormalities, high rates of intrauterine growth restriction, and markedly reduced postnatal survival (Edwards et al., 1960). Among live-born autosomal trisomies, Trisomy 18 is the second most common after Trisomy 21 but carries a substantially poorer prognosis.

3.1 Genetic basis and cytogenetic forms

The majority of Edwards syndrome cases (~90–95%) result from full trisomy 18, arising due to meiotic nondisjunction, predominantly of maternal origin. As with other trisomies, maternal age is a significant risk factor, reflecting age-related meiotic instability in oocytes (Cereda & Carey, 2012). Less commonly, Edwards syndrome may present as mosaic trisomy 18, in which a subset of cells carries the extra chromosome, or as partial trisomy 18, resulting from unbalanced translocations involving chromosome 18. Mosaic and partial forms are typically associated with milder phenotypes and increased survival compared to complete trisomy, although clinical variability remains substantial (Eichenlaub-Ritter, 1996).

3.2 Epidemiology and natural history

The estimated prevalence of Trisomy 18 is approximately one in 5000–6000 live births, although the true incidence is considerably higher due to high rates of spontaneous miscarriage and stillbirth (Inkster et al., 2021). Prenatal studies indicate that only a small proportion of affected fetuses survive to term, reflecting the profound developmental disruptions caused by chromosome 18 gene dosage imbalance. Female predominance among live births has been consistently observed, possibly due to sex-specific differences in fetal viability (Cortezzo et al., 2022). Survival outcomes remain poor. More than 50% of live-born infants die within the first week of life, and fewer than 10% survive beyond the first year. Long-term survival, while rare, has been documented primarily in mosaic cases and is often associated with severe neurodevelopmental impairment and complex medical needs (Reischer et al., 2025).

3.3 Clinical phenotype

Edwards syndrome is characterized by a distinctive constellation of craniofacial, musculoskeletal, cardiovascular, and visceral anomalies. Prenatal growth restriction is a hallmark feature and is detectable on second-trimester ultrasonography. Craniofacial abnormalities include a small, narrow head, prominent occiput, low-set malformed ears, micrognathia, and short palpebral fissures (Bronsteen et al., 2004). Musculoskeletal abnormalities are particularly characteristic and include clenched hands with overlapping fingers, rocker-bottom feet, joint contractures, and hypoplastic nails. These features are considered highly suggestive of Trisomy 18 in the appropriate clinical context. Major internal organ anomalies are common, with congenital heart defects—such as ventricular septal defects, atrial septal defects, and patent ductus arteriosus—present in over 90% of cases (Watson et al., 2008). Renal anomalies, gastrointestinal malformations, and central nervous system defects, including agenesis of the corpus callosum, further contribute to morbidity and mortality.

3.4 Neurodevelopmental and systemic impairment

Neurological involvement in Edwards syndrome is profound and universal among survivors. Severe intellectual disability, hypotonia evolving into hypertonia, feeding difficulties, and respiratory instability are typical. Brain imaging often reveals structural abnormalities reflecting disrupted neurogenesis and neuronal migration during early embryonic development (Klein et al., 2016). These findings are consistent with transcriptomic studies demonstrating dysregulation of genes involved in neuronal differentiation, synaptogenesis, and cell cycle control on chromosome 18. Systemic complications, including recurrent infections, apnea, and failure to thrive, are common and contribute significantly to early mortality. Even with intensive medical care, quality of life is severely compromised, raising complex ethical considerations regarding the extent and goals of intervention (Van Dyke & Allen, 1990).

3.5 Prenatal detection and prognostic Indicators

Prenatal detection of Trisomy 18 has improved substantially with advances in screening and diagnostic technologies. First- and second-trimester ultrasound findings, such as increased nuchal translucency, growth restriction, choroid plexus cysts, and cardiac anomalies, prompt further genetic evaluation. However, definitive diagnosis relies on cytogenetic or molecular confirmation, increasingly guided by noninvasive prenatal screening strategies discussed in later sections of this chapter. Prognosis is influenced by cytogenetic subtype (full vs. mosaic), severity of structural anomalies, and access to specialized neonatal care. Nonetheless, Edwards syndrome remains one of the most life-limiting chromosomal disorders compatible with live birth.

4. Trisomy 13

Patau syndrome is a rare but devastating autososomal aneuploidy caused by the presence of an additional copy of chromosome 13. First described by Klaus Patau and colleagues in 1960, the condition is characterized by severe congenital malformations affecting midline structures, the central nervous system, cardiovascular system, and craniofacial development (Patau et al., 1960). Among the three common autosomal trisomies compatible with live birth, Trisomy 13 is associated with the most severe structural anomalies and the poorest overall prognosis.

4.1 Genetic and cytogenetic mechanisms

Approximately 75%–85% of Patau syndrome cases result from full trisomy 13 due to meiotic nondisjunction, most frequently of maternal origin. As with Trisomy 18 and Trisomy 21, increasing maternal age is a recognized risk factor, reflecting age-related meiotic instability (Babay et al., 2019). In contrast to Edwards syndrome, a relatively higher proportion of Trisomy 13 cases arise from Robertsonian translocations, particularly involving chromosomes 13 and 14. These unbalanced translocations may be inherited from a balanced parental carrier, significantly increasing recurrence risk in affected families (Ahmad et al., 2010). Mosaic trisomy 13, resulting from postzygotic mitotic errors, accounts for a minority of cases and is generally associated with milder phenotypes and extended survival compared to complete trisomy. However, phenotypic variability remains wide and unpredictable, even within mosaic presentations.

4.2 Epidemiology and prenatal loss

The estimated prevalence of Trisomy 13 among live births ranges from one in 10,000 to one in 16,000, though true incidence is substantially higher due to high rates of spontaneous abortion and intrauterine fetal demise (Cook, 2014). Prenatal cytogenetic studies suggest that a large proportion of Trisomy 13 conceptions result in early pregnancy loss, reflecting the essential role of chromosome 13 genes in early embryogenesis and organ patterning. Among live-born infants, survival is extremely limited. Median survival is measured in days to weeks, and fewer than 5%–10% of affected infants survive beyond the first year of life. As observed in Trisomy 18, female predominance among live births has been reported, likely reflecting sex-specific differences in fetal viability (Goel et al., 2019).

4.3 Core clinical features

Patau syndrome is classically associated with midline developmental defects. Craniofacial anomalies are prominent and include holoprosencephaly, microphthalmia or anophthalmia, cleft lip and/or palate, hypotelorism, and malformed or low-set ears (Sankaran & Kyle, 2014, p. 289). Scalp defects, particularly aplasia cutis congenita, are considered highly suggestive of Trisomy 13 in the appropriate clinical context. Central nervous system anomalies are universal and severe, with holoprosencephaly representing the most characteristic finding. These defects are frequently accompanied by profound neurodevelopmental impairment, seizures, and abnormalities of brainstem function. Limb abnormalities such as postaxial polydactyly, clenched hands, and rocker-bottom feet further contribute to the diagnostic phenotype.

4.4 Visceral malformations and systemic involvement

Major internal organ anomalies are present in the majority of affected individuals. Congenital heart disease occurs in over 80% of cases, commonly manifesting as ventricular septal defects, atrial septal defects, and complex conotruncal malformations (Khan et al., 20222). Renal anomalies, including polycystic kidneys and hydronephrosis, are frequently observed, along with gastrointestinal defects such as omphalocele and intestinal malrotation. Endocrine, hematological, and immune dysfunctions have also been reported, although these are often overshadowed by life-threatening structural anomalies (Kravtzova et al., 1975). Feeding difficulties, recurrent infections, and respiratory insufficiency are common contributors to early mortality.

4.5 Neurodevelopmental impact and survival considerations

Survivors of the neonatal period universally exhibit profound intellectual disability, severe motor impairment, and limited sensory responsiveness. Functional independence is not achievable, and long-term survivors require comprehensive medical and supportive care. Advances in neonatal intensive care have led to sporadic reports of extended survival; however, such cases remain exceptional and are frequently associated with mosaic or partial trisomy 13. The severity and multiplicity of anomalies in Patau syndrome raise complex questions regarding the appropriateness and extent of aggressive medical intervention. These considerations are increasingly relevant in the context of early prenatal diagnosis and informed parental decision-making.

4.6 Prenatal detection and diagnostic Clues

Prenatal ultrasound often reveals multiple anomalies suggestive of Trisomy 13, including holoprosencephaly, facial clefts, cardiac defects, and growth restriction. Increased nuchal translucency and abnormal first-trimester biochemical markers may further elevate suspicion. However, definitive diagnosis requires chromosomal or molecular confirmation, increasingly guided by non-invasive prenatal screening approaches discussed later in this chapter.

5. Trisomy 21

Trisomy 21, commonly known as Down syndrome, is the most prevalent and best-characterized autosomal aneuploidy compatible with long-term survival (Bull, 2020). It results from the presence of an extra copy of chromosome 21 and is associated with a distinctive combination of neurodevelopmental, craniofacial, cardiovascular, immunological, and metabolic features. Individuals with Down syndrome frequently survive into adulthood, making it a lifelong condition with evolving clinical and social implications (Desai, 1997).

5.1 Genetic and cytogenetic basis

Approximately 95% of Down syndrome cases arise from full trisomy 21 due to meiotic nondisjunction, with the majority originating from maternal meiosis I errors. Advanced maternal age is the strongest and most consistently identified risk factor, reflecting age-related deterioration of meiotic cohesion and recombination fidelity (Bittles et al., 2007). About 3%–4% of cases are attributable to Robertsonian translocations, most commonly involving chromosomes 14 and 21 or homologous chromosome 21 translocations. These cases are clinically indistinguishable from free trisomy 21 but carry significant implications for familial recurrence risk, particularly when a parent is a balanced translocation carrier (Coppedè, 2016). The remaining 1%–2% of cases present as mosaic trisomy 21, resulting from postzygotic mitotic nondisjunction or trisomy rescue. Mosaic individuals often exhibit milder phenotypes, though clinical variability remains wide and unpredictable (Kazemi et al., 2016).

5.2 Epidemiology and survival trends

The global incidence of Down syndrome is estimated at approximately one in 700–800 live births, although prevalence varies by region, maternal age distribution, and availability of prenatal screening (Chen et al., 2022). Unlike Trisomy 18 and Trisomy 13, survival rates have improved dramatically over the past 5 decades due to advances in neonatal care, surgical correction of congenital anomalies, and comprehensive health surveillance. Current estimates suggest that over 80%–90% of individuals with Down syndrome survive into adulthood, with median life expectancy exceeding 60 years in high-income settings (Al Biltagi, 2015). These improvements have shifted the clinical focus from early mortality to long-term health management, quality of life, and social inclusion.

5.3 Phenotypic spectrum and Core clinical features

Down syndrome is associated with a characteristic but highly variable phenotype. Craniofacial features include brachycephaly, upslanting palpebral fissures, epicanthic folds, flat facial profile, small ears, and a protruding tongue. Generalized hypotonia is common in infancy and contributes to delayed motor development (Liu et al., 2025). Congenital heart disease is present in approximately 40%–50% of affected individuals, with atrioventricular septal defects, ventricular septal defects, and atrial septal defects being the most frequent. Congenital heart disease remains a major determinant of early morbidity and mortality (Buhasan et al., 2025). Gastrointestinal anomalies, including duodenal atresia, Hirschsprung disease, and imperforate anus, occur more frequently than in the general population. Musculoskeletal features such as joint hypermobility, atlantoaxial instability, and short stature further contribute to clinical complexity. Fig. 4.2 is the representative diagram showing the developmental consequences of chromosomal trisomy.

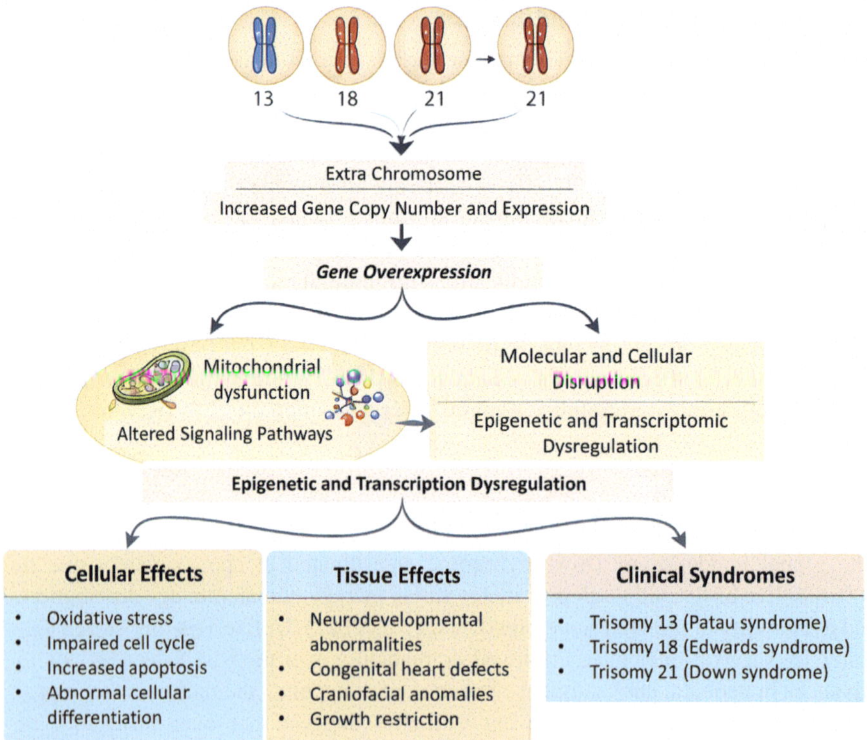

FIGURE 4.2

Developmental consequences of chromosomal trisomy.

5.4 Neurodevelopmental and Cognitive profile

Intellectual disability is a universal feature of Down syndrome, typically ranging from mild to moderate severity. Cognitive impairment primarily affects language acquisition, executive function, working memory, and adaptive behavior, while visuospatial skills and social responsiveness are relatively preserved (Karmiloff-Smith et al., 2016). Structural and functional neuroimaging studies demonstrate reduced total brain volume, altered cortical organization, and delayed myelination, reflecting disrupted neurodevelopmental trajectories. Importantly, individuals with Down syndrome exhibit a markedly increased risk of early-onset Alzheimer disease, driven by triplication of the amyloid precursor protein gene located on chromosome 21. Virtually all adults with Down syndrome develop Alzheimer-type neuropathology by the fourth decade of life, although clinical dementia onset varies considerably (Lott & Head, 2019).

5.5 Systemic comorbidities and medical risks

Down syndrome is associated with a distinctive pattern of systemic comorbidities. Endocrine disorders, particularly hypothyroidism, are common and require lifelong surveillance. Immune dysregulation contributes to increased susceptibility to infections and autoimmune conditions. Hematological abnormalities include a markedly elevated risk of transient abnormal myelopoiesis and acute megakaryoblastic leukemia in childhood (Miller et al., 1983). Metabolic disorders, sleep apnea, hearing and vision impairment, and obesity are frequent in adolescence and adulthood. These conditions necessitate structured, multidisciplinary healthcare approaches extending across the lifespan.

5.6 Genotype–phenotype Correlations and gene dosage effects

The phenotypic manifestations of Down syndrome are driven by gene dosage effects involving dosage-sensitive genes on chromosome 21. Transcriptomic studies demonstrate widespread dysregulation of genes involved in neurogenesis, oxidative stress response, synaptic plasticity, and mitochondrial function (Alldred et al., 2021; Perluigi et al., 2014). Importantly, secondary transcriptional effects extend beyond chromosome 21, resulting in global regulatory network disruption. While no single "critical region" fully explains the Down syndrome phenotype, cumulative effects of multiple overexpressed genes are thought to underlie the syndrome's complexity. This insight has informed emerging therapeutic strategies aimed at modulating downstream pathways rather than targeting individual genes.

6. Diagnostic yield of non-invasive prenatal screening

6.1 Yield defined by sensitivity, specificity, and predictive value

NIPS achieves exceptionally high analytical sensitivity and specificity for common autosomal trisomies. Large-scale systematic reviews and meta-analyses consistently demonstrate sensitivities exceeding 99% for T21 and slightly lower but still high values for T18 and T13 (Rose et al., 2022). From a population-screening perspective, this translates into a high screening yield, as a substantial proportion of affected pregnancies are identified early with minimal false positives. However, diagnostic yield, defined as the probability that a positive result reflects the true fetal karyotype, is best captured by positive predictive value (PPV). PPV is strongly influenced by disease prevalence and biological confounders. As a result, while NIPS has an excellent diagnostic yield for T21, its yield for T18 and particularly T13 is significantly lower in unselected populations (Zhang et al., 2022).

6.2 Trisomy-specific differences in non-invasive yield

For Trisomy 21, PPVs frequently exceed 80%–90% in high-risk cohorts and remain acceptably high even in general obstetric populations. Consequently, NIPS functions almost as a *quasi-diagnostic* test for T21 in certain clinical contexts, although confirmation is still required.

In contrast, Trisomy 18 exhibits moderate PPVs that improve substantially when structural abnormalities are present on ultrasound. For Trisomy 13, PPVs are consistently the lowest, often below 40%, reflecting the combined effects of rarity, placental mosaicism, and early embryonic lethality (Yuan et al., 2023). These disparities highlight that non-invasive diagnostic yield is not uniform across trisomies, and clinical interpretation must be condition-specific.

7. Diagnostic yield of invasive prenatal testing

7.1 Cytogenetic and molecular certainty

Invasive diagnostic procedures—CVS and amniocentesis—provide direct access to fetal genetic material, providing definitive chromosomal analysis through karyotyping, quantitative fluorescent PCR, or chromosomal microarray analysis. The diagnostic yield of invasive testing for autosomal trisomies approaches 100%, limited primarily by technical failure or confined placental mosaicism in CVS samples. Unlike NIPS, invasive testing does not rely on probabilistic inference. A confirmed trisomic karyotype establishes diagnosis unequivocally, allowing clinicians and families to make informed decisions with maximal certainty. This absolute diagnostic yield is particularly critical for T18 and T13, where false-positive NIPS results are more common and prognostic counseling is highly consequential (Badeau et al., 2017). A comparative analysis of NIPS and invasive diagnostic approaches for prenatal detection of trisomies is presented in Table 4.2.

7.2 Procedure-related risk and its impact on yield

Historically, the diagnostic yield of invasive testing was weighed against a perceived high risk of procedure-related miscarriage. Contemporary evidence, however,

Table 4.2 Comparison of non-invasive screening and invasive diagnostic approaches for prenatal detection of trisomies.

Parameter	NIPS	Ultrasound screening	Chorionic villus sampling (CVS)	Amniocentesis
Type	Screening	Screening	Diagnostic	Diagnostic
Sample source	Maternal blood (cfDNA)	Imaging	Placental tissue	Amniotic fluid
Detection timing	≥10 weeks	11–20 weeks	10–13 weeks	≥15 weeks
Sensitivity for T21	>99%	Moderate	~100%	~100%
PPV variability	Yes	High variability	High	High
Risk to fetus	None	None	Low (~0.1–0.3%)	Low (~0.1–0.3%)
Confined placental mosaicism effect	Yes	No	Possible	Rare
Diagnostic certainty	No	No	Yes	Yes

demonstrates that this risk has been substantially overestimated in earlier literature. Modern cohort studies report procedure-related pregnancy loss rates of approximately 0.1%–0.3%, comparable to background miscarriage rates when appropriately controlled (Wei et al., 2023). This reduction in procedural risk fundamentally alters the risk–benefit calculus, particularly when invasive testing is used selectively following high-risk NIPS results rather than as a primary screening tool.

8. Comparative yield in clinical pathways

8.1 Stand-alone NIPS versus combined NIPS–invasive strategy

When used as a stand-alone test, NIPS offers high screening yield but limited diagnostic yield, especially for T18 and T13. In contrast, a combined pathway, in which NIPS serves as a triage tool followed by confirmatory invasive testing, maximizes both yield dimensions. Evidence from multiple healthcare systems shows that this approach reduces invasive testing rates by 40%–70% while preserving near-complete diagnostic ascertainment of true trisomy cases (Badeau et al., 2017). This tiered strategy is now widely regarded as the optimal balance between safety and certainty, particularly in populations with increasing maternal age and expanding access to genomic screening.

8.2 Role of ultrasound in modulating diagnostic yield

Ultrasound acts as a critical modifier of diagnostic yield in both invasive and non-invasive pathways. Structural anomalies dramatically increase the pre-test probability of T18 and T13, thereby improving the PPV of NIPS and strengthening the indication for invasive confirmation. Conversely, normal ultrasound findings in the presence of a positive NIPS result—especially for T13—raise suspicion of a false-positive screen and reinforce the necessity of invasive testing before irreversible decisions are made (Yuan et al., 2023). Thus, diagnostic yield is maximized not by any single modality but by integrated interpretation.

9. Yield comparison in special populations

9.1 Low-risk and general obstetric populations

In low-risk populations, NIPS retains high sensitivity but exhibits lower PPVs due to reduced prevalence. In these settings, invasive testing as a primary strategy would yield unacceptably high procedure-related harm. Therefore, NIPS provides a high screening yield with minimal risk, while invasive testing is reserved for confirmation, preserving overall diagnostic yield at the population level (Marton et al., 2025).

9.2 Clinical and ethical implications of diagnostic yield differences

Differences in diagnostic yield have direct ethical implications. High diagnostic certainty is essential when test results may influence decisions regarding pregnancy continuation or termination, particularly for T18 and T13, which are associated with profound morbidity and limited survival. Reliance on non-invasive screening alone in such contexts risks both false reassurance and unwarranted termination. The literature consistently emphasizes that diagnostic yield, not analytical accuracy alone, should guide clinical pathways. This principle underpins international recommendations that position NIPS as a powerful screening tool but reaffirm invasive testing as indispensable for diagnosis (Badeau et al., 2017).

10. Management and treatment strategies

10.1 Principles guiding management of trisomy disorders

Management of trisomy disorders is fundamentally shaped by three interrelated factors: biological severity, anticipated survival, and family values and goals of care. Trisomy 21, Trisomy 18, and Trisomy 13 differ markedly in natural history, prognosis, and long-term outcomes; consequently, management strategies must be condition-specific rather than uniform. Across all trisomies, however, several overarching principles apply. First, management is supportive rather than curative. The chromosomal basis of trisomy disorders precludes curative genetic therapy at present, and clinical care focuses on optimizing health, function, comfort, and quality of life. Second, management is inherently multidisciplinary, requiring coordination among obstetricians, neonatologists, pediatric subspecialists, geneticists, nurses, rehabilitation professionals, and psychosocial support teams. Third, decision-making must be family-centered and ethically grounded, recognizing parental autonomy while ensuring that counseling is accurate, balanced, and free from coercion.

10.2 Counseling following prenatal diagnosis

Prenatal identification of trisomy profoundly alters the trajectory of pregnancy management. Counseling following diagnosis should be non-directive, longitudinal, and individualized, addressing medical facts, prognostic variability, and parental preferences (Heald et al., 2012). Families should be informed not only of typical outcomes but also of the range of possible clinical scenarios, particularly in mosaic or partial trisomies. For Trisomy 21, counseling increasingly emphasizes long-term survivorship, developmental potential, and quality of life alongside medical comorbidities (Peterson et al., 2024). In contrast, counseling for Trisomy 18 and Trisomy 13 must address the high likelihood of fetal demise or early neonatal death, while acknowledging that survival beyond infancy, though uncommon, is possible. Clear differentiation between screening results and confirmed diagnosis is essential, particularly when NIPS precedes invasive confirmation. Counseling should explicitly address diagnostic uncertainty, especially for Trisomy 13 and Trisomy 18, where false-positive screening results are more frequent.

10.3 Pregnancy continuation or termination

Decisions regarding pregnancy continuation are deeply personal and influenced by medical, cultural, religious, and legal factors. Clinicians must provide accurate prognostic information without framing continuation or termination as the "expected" choice. For Trisomy 18 and Trisomy 13, where prognosis is poor, some families elect palliative-focused care from birth, while others pursue active medical intervention. Both approaches are ethically legitimate when based on informed consent. Prenatal management for continued pregnancies includes enhanced fetal surveillance, serial ultrasound examinations to monitor growth and organ development, and anticipatory planning for delivery and neonatal care. For pregnancies complicated by severe anomalies, perinatal palliative care teams may be involved before birth to support families and coordinate goals of care.

11. Conclusion

NIPS has emerged as an advanced detection method of common autosomal trisomies, particularly Trisomy 21, Trisomy 18, and Trisomy 13. By analyzing cell-free fetal DNA in maternal plasma, NIPS provides highly sensitive and specific early risk assessment while avoiding the procedural risks associated with invasive diagnostic testing. Its implementation has significantly reduced unnecessary invasive procedures and improved prenatal detection rates, especially for Trisomy 21, where screening performance is most robust. However, important biological limitations persist, including the placental origin of circulating fetal DNA, which may result in discordant findings due to confined placental mosaicism or other maternal–fetal factors. These limitations are more pronounced for Trisomy 18 and Trisomy 13, where positive predictive values are lower, emphasizing that NIPS remains a screening rather than a diagnostic test. Confirmatory invasive testing through chorionic villus sampling or amniocentesis is therefore essential following positive results. Optimal clinical practice requires integration of NIPS with ultrasound evaluation, genetic counseling, and confirmatory diagnostics. When applied within a multidisciplinary, ethically grounded framework, NIPS represents a powerful tool for enhancing prenatal care and supporting informed reproductive decision-making.

References

Ahmad, M., Silvera-Redondo, C., & Hamdan Rodríguez, M. (2010). Nondisjunction and chromosomal anomalies. *Revista Salud Uninorte, 26*(1), 117–133.

Akolekar, R., Beta, J., Picciarelli, G., Ogilvie, C., & D'Antonio, F. (2015). Procedure-related risk of miscarriage following amniocentesis and chorionic villus sampling: A systematic review and meta-analysis. *Ultrasound in Obstetrics and Gynecology, 45*(1), 16–26.

Al Biltagi, M. (2015). Epidemiology and prevalence of Down syndrome. In *Down syndrome children-an update* (pp. 3–44). Bentham Science Publishers.

Alldred, M. J., Penikalapati, S. C., Lee, S. H., Heguy, A., Roussos, P., & Ginsberg, S. D. (2021). Profiling basal forebrain cholinergic neurons reveals a molecular basis for vulnerability within the Ts65Dn model of Down syndrome and Alzheimer's disease. *Molecular Neurobiology, 58*(10), 5141–5162.

Antonarakis, S. E., Skotko, B. G., Rafii, M. S., Strydom, A., Pape, S. E., Bianchi, D. W., & Reeves, R. H. (2020). Down syndrome. *Nature Reviews Disease Primers, 6*(1), 9.

Babay, L.É., Horányi, D., Győrffy, B., & Nagy, G. R. (2019). Evidence for the oocyte mosaicism selection model on the origin of Patau syndrome (trisomy 13). *Acta Obstetricia et Gynecologica Scandinavica, 98*(12), 1558–1564.

Badeau, M., Lindsay, C., Blais, J., Nshimyumukiza, L., Takwoingi, Y., Langlois, S., & Rousseau, F. (2017). Genomics-based non-invasive prenatal testing for detection of fetal chromosomal aneuploidy in pregnant women. *Cochrane Database of Systematic Reviews*, (11).

Birchler, J. A., & Veitia, R. A. (2012). Gene balance hypothesis: Connecting issues of dosage sensitivity across biological disciplines. *Proceedings of the National Academy of Sciences, 109*(37), 14746–14753.

Bittles, A. H., Bower, C., Hussain, R., & Glasson, E. J. (2007). The four ages of Down syndrome. *The European Journal of Public Health, 17*(2), 221–225.

Bronsteen, R., Lee, W., Vettraino, I. M., Huang, R., & Comstock, C. H. (2004). Second-trimester sonography and trisomy 18. *Journal of Ultrasound in Medicine, 23*(2), 233–240.

Buhasan, A., Al Zayer, L., Al-Jabery, L., Mohamed, A., Almadhoob, A., Zaman, E., & Jeddy, R. (2025). Prevalence of congenital anomalies and predictive factors in pregnant women in Bahrain: A retrospective analysis. *Life, 15*(4), 650.

Bull, M. J. (2020). Down syndrome. *New England Journal of Medicine, 382*(24), 2344–2352.

Carey, J. C. (2021). *Trisomy 18 and trisomy 13 syndromes*. Cassidy and Allanson's management of genetic syndromes.

Cereda, A., & Carey, J. C. (2012). The trisomy 18 syndrome. *Orphanet Journal of Rare Diseases, 7*(1), 81.

Chen, L., Wang, L., Wang, Y., Hu, H., Zhan, Y., Zeng, Z., & Liu, L. (2022). Global, regional, and national burden and trends of Down syndrome from 1990 to 2019. *Frontiers in Genetics, 13*, Article 908482.

Cook, J. P. (2014). *Investigating clustering in trisomy 18 and trisomy 13*. Queen Mary University of London. Doctoral dissertation.

Coppedè, F. (2016). Risk factors for Down syndrome. *Archives of Toxicology, 90*(12), 2917–2929.

Cortezzo, D. E., Tolusso, L. K., & Swarr, D. T. (2022). Perinatal outcomes of fetuses and infants diagnosed with trisomy 13 or trisomy 18. *The Journal of Pediatrics, 247*, 116–123.

De Jong, A., Dondorp, W. J., de Die-Smulders, C. E., Frints, S. G., & De Wert, G. M. (2010). Non-invasive prenatal testing: Ethical issues explored. *European Journal of Human Genetics, 18*(3), 272–277.

Desai, S. S. (1997). Down syndrome: A review of the literature. *Oral Surgery, Oral Medicine, Oral Pathology, Oral Radiology, and Endodontology, 84*(3), 279–285.

Edwards, J. H., Harnden, D. G., Cameron, A. H., Crosse, V. M., & Wolf, O. H. (1960). A new trisomic syndrome. *The lancet, 275*(7128), 787–790.

Eichenlaub-Ritter, U. (1996). Parental age-related aneuploidy in human germ cells and offspring: A story of past and present. *Environmental and Molecular Mutagenesis, 28*(3), 211–236.

Goel, N., Morris, J. K., Tucker, D., de Walle, H. E., Bakker, M. K., Kancherla, V., & Morgan, M. (2019). Trisomy 13 and 18—Prevalence and mortality—A multi-registry population based analysis. *American Journal of Medical Genetics, Part A, 179*(12), 2382–2392.

Hassold, T., & Hunt, P. (2001). To err (meiotically) is human: The genesis of human aneuploidy. *Nature Reviews Genetics, 2*(4), 280–291.

Heald, B., Melvin, A., & Madeo, A. C. (2012). Presented abstracts from the thirty first annual education conference of the national society of genetic counselors (Boston, MA, October 2012). *Journal of Genetic Counseling, 21*(6), 884–984.

Inkster, A. M., Fernández-Boyano, I., & Robinson, W. P. (2021). Sex differences are here to stay: Relevance to prenatal care. *Journal of Clinical Medicine, 10*(13), 3000.

Jones, K. T., & Lane, S. I. (2013). Molecular causes of aneuploidy in mammalian eggs. *Development, 140*(18), 3719–3730.

Karmiloff-Smith, A., Al-Janabi, T., D'souza, H., Groet, J., Massand, E., Mok, K., & Strydom, A. (2016). The importance of understanding individual differences in Down syndrome. *F1000Research, 5*. F1000-Faculty.

Kazemi, M., Salehi, M., & Kheirollahi, M. (2016). Down syndrome: Current status, challenges and future perspectives. *International journal of molecular and cellular medicine, 5*(3), 125.

Khan, U., Hussain, A., Usman, M., & ul Abiddin, Z. (2022). An infant with Patau syndrome associated with congenital heart defects. *Annals of Medicine and Surgery, 80*, Article 104100.

Klein, J. L., Lemmon, M. E., Northington, F. J., Boltshauser, E., Huisman, T. A., & Poretti, A. (2016). Clinical and neuroimaging features as diagnostic guides in neonatal neurology diseases with cerebellar involvement. *Cerebellum & Ataxias, 3*(1), 1.

Kravtzova, G. I., Lazjuk, G. I., & Lurie, I. W. (1975). The malformations of the urinary system in autosomal disorders. *Virchows Archiv, 368*(2), 167–178.

Lamb, N. E., Sherman, S. L., & Hassold, T. J. (2005). Effect of meiotic recombination on the production of aneuploid gametes in humans. *Cytogenetic and Genome Research, 111* (3–4), 250–255.

Letourneau, A., Santoni, F. A., Bonilla, X., Sailani, M. R., Gonzalez, D., Kind, J., & Antonarakis, S. E. (2014). Domains of genome-wide gene expression dysregulation in Down's syndrome. *Nature, 508*(7496), 345–350.

Levy, B., Hoffmann, E. R., McCoy, R. C., & Grati, F. R. (2021). Chromosomal mosaicism: Origins and clinical implications in preimplantation and prenatal diagnosis. *Prenatal Diagnosis, 41*(5), 631–641.

Liu, J., Wang, S., Zhou, S., Ma, D., Hu, L., Huang, H., & He, J. (2025). Cost-effectiveness of different screening strategies for Down syndrome: A real-world analysis in 140,472 women. *Frontiers in Public Health, 13*, Article 1535381.

Lott, I. T., & Head, E. (2019). Dementia in Down syndrome: Unique insights for Alzheimer disease research. *Nature Reviews Neurology, 15*(3), 135–147.

Marton, T., Erdélyi, Z. R., Takai, M., Mészáros, B., Supák, D., Ács, N., & Valent, S. (2025). Systematic review of accuracy differences in NIPT methods for common aneuploidy screening. *Journal of Clinical Medicine, 14*(8), 2813.

Mehler, M. F., & Mattick, J. S. (2007). Noncoding RNAs and RNA editing in brain development, functional diversification, and neurological disease. *Physiological Reviews, 87*(3), 799–823.

Mendioroz, M., Do, C., Jiang, X., Liu, C., Darbary, H. K., Lang, C. F., & Tycko, B. (2015). Trans effects of chromosome aneuploidies on DNA methylation patterns in human Down syndrome and mouse models. *Genome Biology, 16*(1), 263.

Miller, M., Cosgriff, J. M., & Opitz, J. M. (1983). Hematological abnormalities in newborn infants with Down syndrome. *American Journal of Medical Genetics, 16*(2), 173–177.

Morris, J. K., Alberman, E., Scott, C., & Jacobs, P. (2008). Is the prevalence of Klinefelter syndrome increasing? *European Journal of Human Genetics, 16*(2), 163–170.

Patau, K., Smith, D., Therman, E., Inhorn, S., & Wagner, H. (1960). Multiple congenital anomaly caused by an extra autosome. *The Lancet, 275*(7128), 790–793.

Perluigi, M., Di Domenico, F., & Buttterfield, D. A. (2014). Unraveling the complexity of neurodegeneration in brains of subjects with Down syndrome: Insights from proteomics. *Proteomics - Clinical Applications, 8*(1–2), 73–85.

Peterson, J. K., Clarke, S., Gelb, B. D., Kasparian, N. A., Kazazian, V., Pieciak, K., & American Heart Association Pediatric Cardiovascular Nursing Committee of the Council on Cardiovascular and Stroke Nursing; Council on Clinical Cardiology; Council on Genomic and Precision Medicine; and Council on Cardiovascular Radiology and Intervention. (2024). Trisomy 21 and congenital heart disease: Impact on health and functional outcomes from birth through adolescence: A scientific statement from the American heart association. *Journal of the American Heart Association, 13*(19), Article e036214.

Rasmussen, S. A., Wong, L. Y. C., Yang, Q., May, K. M., & Friedman, J. M. (2003). Population-based analyses of mortality in trisomy 13 and trisomy 18. *Pediatrics, 111* (4), 777–784.

Reischer, T., Dorittke, T., Klebermaß-Schrehof, K., Berger, A., Dremsek, P., Ragossnig, J., & Yerlikaya-Schatten, G. (2025). Clinical course and perinatal management of fetuses and newborns affected by trisomy 13 and 18: A retrospective single-centre cohort study. *European Journal of Pediatrics, 185*(1), 30.

Robinson, W. P. (2000). Mechanisms leading to uniparental disomy and their clinical consequences. *BioEssays, 22*(5), 452–459.

Rose, N. C., Barrie, E. S., Malinowski, J., Jenkins, G. P., McClain, M. R., LaGrave, D., & ACMG Professional Practice and Guidelines Committee. (2022). Systematic evidence-based review: The application of noninvasive prenatal screening using cell-free DNA in general-risk pregnancies. *Genetics in Medicine, 24*(7), 1379–1391.

Sankaran, S., & Kyle, P. (2014). *Abnormalities of the face and neck.* Twining's textbook of fetal abnormalities E-Book: Twining's textbook of fetal abnormalities E-Book.

Sherman, S. L., Allen, E. G., Bean, L. H., & Freeman, S. B. (2007). Epidemiology of Down syndrome. *Mental Retardation and Developmental Disabilities Research Reviews, 13*(3), 221–227.

Springett, A. L., & Morris, J. K. (2014). Antenatal detection of Edwards (trisomy 18) and Patau (trisomy 13) syndrome: England and Wales 2005-2012. *Journal of Medical Screening, 21*(3), 113–119.

Srinivasan, K., Canarte, C., Valientes, S. D. A., Sanchez-Lara, P. A., & Langston, S. J. (2025). Updates in trisomy 18. *NeoReviews, 26*(12), e820–e834.

Van Dyke, D. C., & Allen, M. (1990). Clinical management considerations in long-term survivors with trisomy 18. *Pediatrics, 85*(5), 753–759.

van't Oever, R. M., Verweij, E. J. T., & de Haas, M. (2025). How I use noninvasive prenatal testing for red blood cell and platelet antigens. *Blood, 145*(20), 2266–2274.

Volk, M., Maver, A., Hodžić, A., Lovrečić, L., & Peterlin, B. (2017). Transcriptome profiling uncovers potential common mechanisms in fetal trisomies 18 and 21. *OMICS: A Journal of Integrative Biology, 21*(10), 565–570.

Watson, W. J., Miller, R. C., Wax, J. R., Hansen, W. F., Yamamura, Y., & Polzin, W. J. (2008). Sonographic findings of trisomy 18 in the second trimester of pregnancy. *Journal of Ultrasound in Medicine, 27*(7), 1033–1038.

Webster, A., & Schuh, M. (2019). Mechanisms of aneuploidy in human eggs. *Trends in Cell Biology, 27*(1), 55–68.

Wei, R., Li, J., Xia, Y., Wang, C., Lu, X., Fang, Y., & Zhu, J. (2023). Application of non-invasive prenatal testing to 91,280 spontaneous pregnancies and 3477 pregnancies conceived by in vitro fertilization. *Molecular Cytogenetics, 16*(1), 25.

Wu, J., Springett, A., & Morris, J. K. (2013). Survival of trisomy 18 (Edwards syndrome) and trisomy 13 (Patau Syndrome) in England and Wales: 2004–2011. *American Journal of Medical Genetics, Part A, 161*(10), 2512–2518.

Yuan, X., Yong, W., Dai, L., Wang, W., & Wu, L. (2023). The role of non-invasive prenatal testing and ultrasound in prenatal screening of fetal chromosomal abnormalities in singleton: A retrospective study. *Annals of Translational Medicine, 11*(2), 111.

Zhang, Y., Xu, H., Zhang, W., & Liu, K. (2022). Non-invasive prenatal testing for the detection of trisomy 13, 18, and 21 and sex chromosome aneuploidies in 68,763 cases. *Frontiers in Genetics, 13*, Article 864076.

CHAPTER

5

Sex chromosome aneuploidies: Klinefelter syndrome, triple X syndrome, and monosomy X (turner syndrome)

Insha Bashir

Department of Biochemistry, GMC, Srinagar, University of Kashmir, J&K, India

1. Introduction

One of the most prevalent types of chromosomal abnormalities in the human population is sex chromosome aneuploidy (SCA). A variety of clinical characteristics, such as decreased fertility or infertility, poor reproductive function, delays in speech and language acquisition, and abnormalities in motor development, are frequently linked to these diseases. For the evaluation of fetal chromosomal abnormalities, non-invasive prenatal screening (NIPS) has gained popularity in recent years. This method identifies fetal sex as well as detects sex chromosomal abnormalities during pregnancy (Deng et al., 2021). A significant class of genetic disorders seen in prenatal medicine is known as SCAs, which are brought on by numerical anomalies involving the X or Y chromosome. The most frequently recognized of them are monosomy X (45, X), also referred to as Turner syndrome, Triple X syndrome (47, XXX), and Klinefelter syndrome (47, XXY). In contrast to many autosomal aneuploidies, these disorders cause a wide variety of clinical symptoms, from minor developmental abnormalities to serious reproductive, endocrine, and cardiovascular problems (Groth et al., 2013). A significant percentage of cases go undetected without genetic testing because phenotypic traits are often moderate or ambiguous after birth. The study of cell-free fetal DNA (cffDNA) in maternal plasma has reformed the prenatal risk assessment for chromosomal disorders. NIPS was first used to identify common autosomal trisomies, but it has since been progressively extended to screen sex chromosomal aneuploidies and common genetic disorders (Bianchi & Wilkins-Haug, 2014; Norton et al., 2015). High sensitivity is provided by this technology, which also eliminates the need for invasive procedures like amniocentesis and chorionic villus sampling. Notwithstanding these benefits, there are certain biological and interpretative difficulties with NIPS screening for SCAs. Variable fetal fraction, vanishing twin pregnancies, maternal sex chromosomal mosaicism, and limited placental mosaicism are some of the factors that can affect

Non-Invasive Prenatal Screening of Rare Fetal Genetic Diseases. https://doi.org/10.1016/B978-0-443-27660-6.00010-9

test accuracy and provide inconsistent results (Grati, 2014; Wang et al., 2014). Furthermore, because of the phenotypic diversity, it is challenging to determine the clinical relevance of a positive screening result for diseases like Klinefelter or Triple X syndrome during pregnancy. Because Turner syndrome is linked to fetal hydrops, congenital abnormalities, and an increased risk of miscarriage, it has special clinical significance in prenatal screening (Gravholt et al., 2023).In the context of screening fetal genetic disorders, this chapter examines the function of NIPS in the identification of Klinefelter syndrome, Triple X syndrome, and monosomy X. The genetic foundation, epidemiology, clinical implications, and limits of NIPS for various illnesses are highlighted, along with counseling and confirmatory diagnostic testing concerns. The proper clinical use of NIPS in modern prenatal care requires a thorough comprehension of these elements.

2. Principles of non-invasive prenatal screening

The foundation of NIPS is the examination of brief segments of cell-free DNA that circulate in maternal plasma, some of which come from placental trophoblasts and represent the fetal genome (Bianchi et al., 2012; Lo et al., 1997) (Fig. 5.1). The majority of the cell-free DNA found in pregnant women's blood is maternal in origin. Merely 10% or less comes from the fetus, or more specifically, the placenta. For proper analysis, the "fetal fraction" of cell-free DNA in the mother's blood must be higher than the minimal level required by NIPT; most labs put this limit at, say, 4% (Canick et al., 2013). If testing is conducted before nine or 10 weeks of pregnancy, the fraction of cffDNA in the mother's blood may not be sufficiently substantial, even though it can be detected as early as 4 weeks. As a result, earlier testing could produce an unreliable or unsuccessful result. The fetal fraction may still be very low in subsequent testing, though, because of maternal variables that require

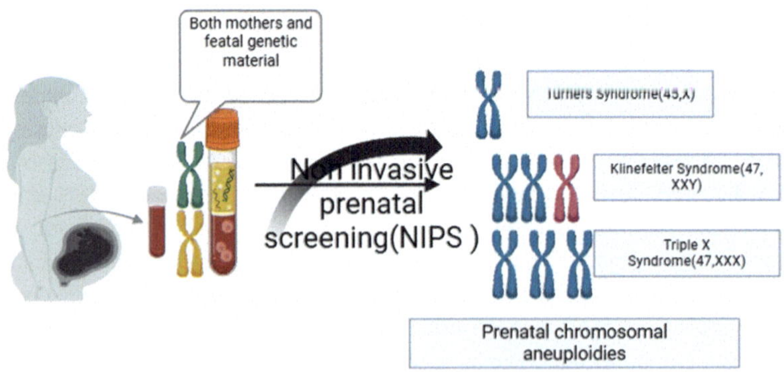

FIGURE 5.1

Diagram illustrating the work flow of non-invasive prenatal screening with regard to sex chromosome aneuploidy.

more research. enhanced maternal body weight due to a dilution effect and likely enhanced adipocyte turnover in obese women is one well-established risk factor for a poor outcome (Ashoor et al., 2013; Wang et al., 2013). By employing genome-wide or focused next-generation sequencing techniques, NIPS measures chromosomal representation and makes it possible to identify relative chromosomal gains or losses as early as the first trimester of pregnancy. NIPS is a screening tool, not a diagnostic one, even with its excellent analytical sensitivity. The reported failure rates, which range from 0% to 5%, differ significantly amongst laboratories (Benn et al., 2013). Repeating the NIPT or looking at other testing options after a failed result can be done, but this adds time to the screening and diagnosis procedure, which could influence the woman's ability or choice to have a non-surgical termination. Additional data about failure rates and risk factors for unsuccessful NIPT is required (Huang et al., 2014). Although abnormal results suggest a higher likelihood of a chromosomal abnormality, they must be confirmed by invasive diagnostic techniques such as amniocentesis or chorionic villus collection (ACOG, 2020). The interpretation of sex chromosome aneuploidies is confounded by biological variables such as maternal chromosomal mosaicism, limited placental mosaicism, and technical restrictions that make it difficult to discern minute differences in sex chromosome dosage (Benn et al., 2015).

3. Klinefelter syndrome (47, XXY)

3.1 Genetic background and frequency

Klinefelter syndrome affects males and is represented by the presence of two or more X chromosomes (Bonomi et al., 2017). In 1959, the genetic cause, supernumerary X chromosomes, usually 47, XXY, was discovered (Jacobs & Strong, 1959). Klinefelter Syndrome is one of the most prevalent SCAs, with a frequency rate of between 1/660 to 1/450 in male fetuses born alive (Bojesen et al., 2003; Lanfranco et al., 2004). Klinefelter syndrome is associated with the karyotype 47, XXY. Rather than being inherited, the extra X chromosome develops irregularly during gametogenesis due to meiotic nondisjunction. Other additional X chromosomes cause testicular fibrosis, hypofunction, and hyalinization, which results in genital problems, most frequently infertility and hypogonadism (Belling et al., 2017).

3.2 Clinical characteristics

The American doctor, Dr. Harry Klinefelter, was the first to identify the clinical characteristic in 1942 (Bonomi et al., 2017). Those who are affected frequently exhibit azoospermia, gynecomastia, undersized testes, and tall stature. Infertility, growth retardation, and aberrant teenage development are the main reasons why patients are referred to hospitals (Leggett et al., 2010; Tartaglia, Cordeiro, et al., 2010). There are many different clinical manifestations of Klinefelter syndrome; some patients experience moderate symptoms or none at all, while others experience severe symptoms such as mental retardation, oligo-atheno-spermia, and hypogonadism (Ottesen et al., 2010). Language challenges, neurodevelopmental abnormalities, and other associated secondary traits such as pes planus, fifth-finger clinodactyly, and elongated extremities (Samango-Sprouse & Gropman, 2013) exist in these individuals. Klinefelter syndrome causes growth retardation, hypotonia, and mildly asymmetrical motor patterns (Samango-Sprouse & Gropman, 2016). Many Klinefelter Syndrome patients go undetected because fetuses frequently lack the usual prenatal ultrasound symptoms, and some affected boys appear asymptomatic at birth. However, replacement medication is required starting at age 11 or 12 due to low testosterone production, which may be conducive to the neurocognitive development of adolescents with Klinefelter syndrome (Samango-Sprouse et al., 2013). Young patients' prognosis and course of therapy will therefore benefit from an early diagnosis. In the middle to late 20th century, neurocognitive differences started to be more widely acknowledged (Daly, 1969). Effective therapeutic management is frequently supported by cognitive, adaptive, and androgen replacement therapy (Ross et al., 2017). However, because of delayed diagnosis, a lack of established treatment regimens, and restricted access to reasonably priced medicines, clinical care is still variable. Fig. 5.2 shows a typical Klinefelter's syndrome karyotype.

3.3 NIPS detection, performance, and interpretation

The use of NIPT for prenatal screening of fetal chromosomal aneuploidy has grown significantly in a number of nations since Lo et al. (1997) discovered cffDNA in maternal blood. NIPT is mostly utilized in clinical practice to test for aneuploid fetal chromosomes 21, 18, and 13, as well as occasionally SCAs (Bianchi & Wilkins-Haug, 2014). While chromosome 21, 18, and 13 aneuploidies are thought to yield very reliable NIPT results, prospective studies have consistently found slightly poorer sensitivities and specificities for SCAs (Song et al., 2013). It may become possible to identify more and more fetuses with Klinefelter Syndrome during the

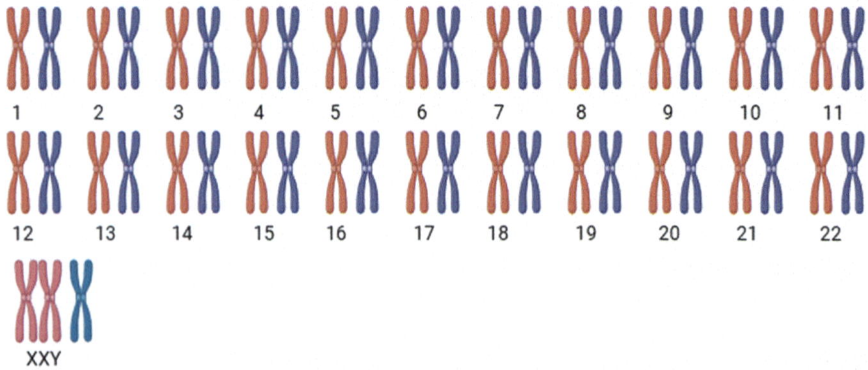

FIGURE 5.2

Karyotype depicting Klinefelter syndromes.

prenatal period as NIPT is widely used. In NIPS, a higher proportion of X chromosomal sequences than Y chromosome material suggests Klinefelter syndrome. Despite reasonable detection rates, positive predictive values are still below those found for common autosomal trisomies (Wang et al., 2014). Constrained placental mosaicism, maternal sex chromosomal mosaicism, and disappearing twin occurrences can all result in false-positive test findings. When considering screening choices, it is imperative to adopt a patient-centered approach. It is advisable to urge patients to contemplate the possible implications of the results for their pregnancy and whether being aware of the outcome would influence their choices or planning (Belling et al., 2017). Zhang et al. (2017) reported a PPV of 77.78% (7/9) for 47, XXY. The overall PPV for SCAs was 54.54%, and NIPT performance was higher in high-risk women (PPV 77.78%) compared to intermediate-risk women. Guo et al. (2022) identified 33 confirmed cases of Klinefelter syndrome among 96 confirmed SCAs, with an overall SCA PPV of 36.9%. Dao et al. (2023) reported a PPV of 50.21% for 47, XXY, with an average combined SCA PPV of 46.08%, noting variability in interpretation across studies. Bussolaro et al. (2023), including 1,531,240 women, demonstrated a pooled PPV of 67.6% (95% CI, 62.5%–72.5%) for XXY. The pooled sensitivity and specificity for SCAs were 94.1% and 99.5%, respectively, with an overall PPV of 49.4%. Lu et al. (2021) reported a PPV of 66.67% for 47, XXY, increasing to 78.95% in advanced maternal age (AMA) women. The frequency of SCAs, particularly 47, XXY, was significantly correlated with maternal age, and AMA was suggested as a risk factor. Kim et al. (2024) reported 8 cases of Klinefelter syndrome with a PPV of 71.43%. NIPT showed significantly higher accuracy for sex chromosome trisomies than monosomy X ($P = .002$). No significant association was found between maternal age and overall fetal SCA incidence. Yang et al. (2021) reported a PPV of 81.13% for 47, XYY/47, XXY in a cohort of 47,800 pregnancies and concluded that NIPT performed better for sex chromosome trisomies than monosomy X. Across studies, NIPT consistently demonstrated higher predictive value for 47, XXY compared to monosomy X, with improved performance in high-risk and AMA groups.

4. Monosomy X (turner syndrome)

4.1 Genetic background and frequency

A sex chromosome condition linked to karyotype heterogeneity is Turner Syndrome (TS). It is frequently linked to the 45, X karyotype and can be associated with severe prenatal abnormalities. Most TS fetuses do not exhibit any overt phenotype, which delays diagnosis and treatment (Bedei et al., 2023). A significant proportion of instances of TS, which affects roughly one in 1700–2000 live-born females, go undetected because of the modest or unusual presentation (Berglund et al., 2020). Furthermore, delayed diagnosis presents a substantial obstacle; the average age of diagnosis for individuals with 45, X karyotypes is 15.1 years, whereas the average age for those with non-45, X karyotypes is 19 years (Berglund et al., 2019).

Significant karyotypic variability is a characteristic of TS, ranging from the classic 45, X karyotype (found in approximately 40%–60%) to mosaic karyotypes involving two or more cell lines and karyotypes with structural X or Y chromosome abnormalities (Gravholt et al., 2024). Both prenatally and postnatally, certain karyotype–phenotype relationships have been found, and these mostly represent the extent of X chromosomal material loss (Fiot et al., 2019; Gravholt et al., 2023; Snyder et al., 2021). A rare type of TS mosaicism, 45, X/47, XXX, affects 1.7 out of all patients with the condition (Kleczkowska et al., 1990). Fig. 5.3 presents the TS karyotype.

4.2 Clinical characteristics

Henry Turner first identified this syndrome in 1938. He noted that the illness was characterized by small height and primary hypergonadotropic hypogonadism (Clemente et al., 2022). The TS prenatal phenotype includes fetal hydrops, elevated NT, and congenital heart problems, which primarily consist of kidney abnormalities, tiny omphaloceles, and left outflow tract anomalies (Bedei et al., 2023). However, studies from Denmark show that only 42% of fetuses with TS are diagnosed prenatally, and many fetuses will present with a normal ultrasound (Bedei et al., 2023).A variety of characteristics make up the postnatal phenotype, ranging from milder phenotypes with few or no obvious clinical symptoms following birth to more common stigmata such as edema of the hands and feet, pterygium colli, short stature, gonadal dysgenesis, and cardiovascular abnormalities (Gravholt et al., 2024). Congenital cardiovascular thyroid autoimmune abnormalities (Mohamed et al., 2018), abnormalities (Essandoh et al., 2014), liver abnormalities (Calanchini et al., 2018), hypertension (Gravholt et al., 1998), hearing loss (Bois et al., 2018), and renal malformations (Hamza et al., 2016) are other main clinical characteristics of TS. Individuals with TS have a lower life expectancy than the general population, with a difference of up to 12.5 years, according to prior studies (Price et al., 1986).

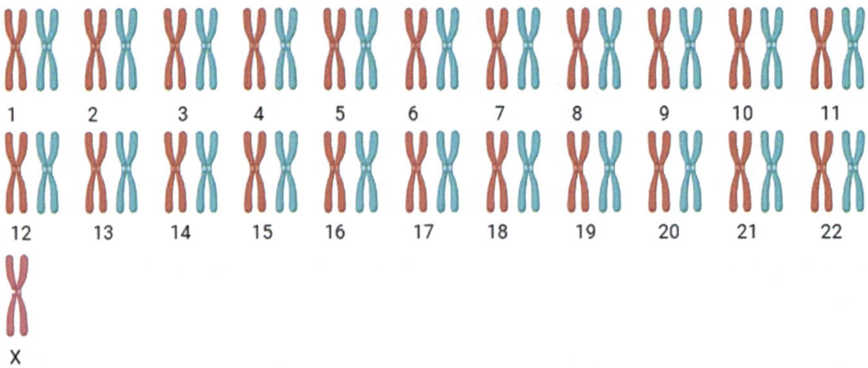

FIGURE 5.3

Karyotype depicting Turner Syndrome.

4.3 NIPS detection, performance, and interpretation

Lau et al. (2014) examined 1982 consecutive patients, 11 of whom had positive SCA tests, and only 85.7% of individuals with NIPT-suspected SCA were fetal in origin. This chromosomal aberration is not ddusually linked to structural abnormalities or abnormal maternal serum screening profiles, yet it may manifest as an abnormal maternal screening test result or abnormal ultrasound finding. TS is very difficult to identify during pregnancy, and the majority of instances are discovered by chance or after birth. Due in large part to limited placental mosaicism, maternal age-related loss of one X chromosome, and technical constraints in sequencing-based dosage assessment, positive predictive values are rather modest (Pe'er Dar et al., 2014). However, when paired with aberrant ultrasound results, NIPS may offer therapeutically valuable early risk assessment.

Zhang et al. (2017) reported a PPV of 29.41% (5/17) for 45, X, with an overall SCA PPV of 54.54%. A PPV of 21% for 45, X, and additionally identification of 15 cases of mosaic TS and other structural abnormalities have been reported previously (Guo et al., 2022), whereas Dao et al. (2023) reported a PPV of 26.05% for 45, X, identifying it as the least accurately detected SCA. Further, it has been reported by Bussolaro et al. (2023) that a pooled PPV of 32.0% (95% CI, 27.0%–37.3%) for monosomy X, significantly lower than for trisomies. Although pooled sensitivity and specificity were high, the overall PPV remained approximately 50% for SCAs and lowest for monosomy X. Lu et al. (2021) reported a PPV of 12.5% for 45, X, increasing to 23.81% in AMA women. NIPT performed better for trisomies than monosomy X, and termination rates were higher in 45, X cases. A recent research by Kim et al. (2024) reported 20 Turner cases with a PPV of 18.75% and confirmed significantly lower accuracy for monosomy X compared to trisomies ($P = .002$). No significant correlation was observed between maternal age and overall SCA incidence. Whereas Yang et al. (2021) reported a PPV of 30.00% for 45, X, and detected additional cases of sex chromosome mosaicism and CNVs incidentally. Across all studies, NIPT consistently showed lower predictive value for monosomy X compared with sex chromosome trisomies. Confined placental mosaicism and biological variability likely contribute to reduced PPV. All studies emphasized the necessity of confirmatory invasive testing and comprehensive genetic counseling.

5. Triple X syndrome (47, XXX)

5.1 Genetic background and frequency

The sex chromosome aneuploidy known as TripleX (47, XXX) syndrome is brought on by an additional X chromosome in females. About one in 1000 female babies are born with this condition, which is the most prevalent sex chromosomal abnormality in females. It is mostly caused by a rare mistake in maternal nondisjunction during meiosis (Slutsky et al., 2023) and possesses a diverse range of physical and mental characteristics. Non-invasive prenatal genetic testing is contributing to an increase in prenatal diagnosis rates of 47, XXX. Previous research indicates that girls with prenatal diagnoses have superior neurodevelopmental outcomes (Wigby et al., 2016).

5.2 Clinical characteristics

Trisomy X individuals present with a variety of phenotypes, such as neuropsychiatric symptoms and dysmorphic traits. There may also be genitourinary abnormalities, seizures, and early ovarian failure. The majority of females with Triple X syndrome develop physically and reproductively normally (Slutsky et al., 2023). Clinical characteristics that may be present include mild hypotonia, modest craniofacial abnormalities, and tall stature. Neurodevelopmental results vary; some people have emotional and behavioral issues, learning disabilities, or speech delays, while others show no symptoms at all. Predicting postnatal outcomes is limited by this diversity, which also makes prenatal counseling more difficult (Tartaglia, Howell, et al., 2010). According to earlier research, females with Triple X syndrome diagnosed during pregnancy had better neurodevelopmental outcomes in terms of IQ and the prevalence of mental health disorders or academic challenges (Robinson et al., 1992).

5.3 NIPS detection, performance, and interpretation

Amniocentesis or the fetal karyotype can identify triple X syndrome during pregnancy. Fetal testing cannot predict when symptoms will appear or how severe they will be. Triple X syndrome is detected during pregnancy as part of routine aneuploidy screening or testing for other fetal chromosome abnormalities; the diagnosis can be surprising (Gross et al., 2022). By identifying excess X chromosomal material, NIPS is able to identify Triple X syndrome. The relatively mild and varied phenotype raises concerns about the clinical value of prenatal identification, even when analytical detection is possible (Benn et al., 2015). Abnormal screening results should be followed by confirmed invasive testing and thorough genetic counseling, much like with other sex chromosomal aneuploidies. It has been identified by Guo et al. (2022) that 17 confirmed cases of 47, XXX among SCA-positive results. Dao et al. (2023) reported a PPV of 44.82% for 47, XXX. Whereas Bussolaro et al. (2023) demonstrated a pooled PPV of 57.5% (95% CI, 51.7%–63.1%) for XXX, with higher PPVs observed for SCAs involving a supernumerary Y chromosome compared with monosomy X. Also, Lu et al. (2021) reported a PPV of 51.72% for 47, XXX, increasing to 53.33% in AMA women. The frequency of 47, XXX was significantly correlated with maternal age. Kim et al. (2024) reported 12 triple X cases with a PPV of 88.89%, and confirmed better performance for trisomies than monosomy X ($P = .002$). Yang et al. (2021) reported a PPV of 70.58% for 47, XXX. Overall, NIPT demonstrated moderate to high predictive performance for 47, XXX, though variability between studies was observed. Maternal age showed an association in some cohorts. Confirmatory testing remained essential.

6. Performance and limitations of NIPS for sex chromosome aneuploidies

The positive predictive value for SCA is lower compared to aneuploidies (Li et al., 2024). Not all genetic disorders are screened for by NIPT. Prenatal diagnostic tests for the particular genetic indication should be taken into consideration, and genetic counseling is advised if a genetic problem is suspected based on ultrasound results, reproductive carrier screening, or family history (Poulton & Hui, 2025). There is still a chance that a fetal chromosomal abnormality could exist but go undetected by NIPT because it is a screening test rather than a diagnostic one (Lindquist et al., 2018).

The existence of a euploid fetus may result in an aberrant maternal plasma cfDNA profile due to genetic variations between the fetus and placenta. In about 1% of pregnancies, confined placental mosaicism develops. Constrained placental mosaicism may cause differences between NIPT results and the actual fetal karyotype because the placenta is the source of pregnancy-derived cfDNA in maternal blood (Mardy & Wapner, 2016). Chromosome variants are prevalent in tumors, and their cfDNA can interfere with NIPT, producing odd outcomes like multiple aneuploidies (Lannoo et al., 2021).

7. Ethical and counseling considerations

The practice of prenatal screening for fetal anomalies is morally sensitive due to its association with (selective) abortion (Clarke, 1997). First of all, if the program's effectiveness is perceived to be based on the percentage of abnormally terminated fetuses, this could subtly encourage women to request an abortion if the fetus is discovered to be impacted. Hence, abortion choices would become a tool for public health. Second, the goal exposes the technique to the so-called "disability rights" or "expressivist critique," which holds that prenatal screening conveys a discriminating message about the value of the lives of those who have the relevant conditions (Parens & Asch, 2003). To steer clear of these moral problems, pertinent policy documents emphasize that the goal of prenatal screening for fetal abnormalities is to empower pregnant women and their partners to make autonomous reproductive decisions rather than to prevent the birth of children with particular abnormalities (Comité Consultatif National d'Éthique, 2013). Since maximizing reproductive liberty as such cannot possibly be a defensible public health goal, this may need to be defined as referring to meaningful choices relating to significant health issues (De Jong & De Wert, 2015).

8. Conclusion

NIPS provides a safe and early approach to risk assessment, but there are ethical, technical, and biological issues with applying it to SCAs. Clinicians, researchers, and policymakers must have a sophisticated grasp of the genetic pathways, clinical variability, and screening constraints. The prudent integration of NIPS with confirmatory assays and excellent genetic counseling will continue to be essential to evidence-based prenatal care as prenatal genomic technologies advance. Because of its low false-positive rate and great sensitivity, NIPT is a first-line screening test for fetal aneuploidies. To guarantee accuracy, invasive diagnostic testing should still be performed to validate a high probability screening result. NIPT has limits that should be shared with patients when doing these tests, despite the fact that it can detect a variety of illnesses. To provide the best care, clinicians should be well-informed about the capabilities, breadth, and cost of the particular NIPT platform they are using.

References

ACOG Publications. (2020). Obstetrics & Gynecology, 136(2), 433–434. https://doi.org/10.1097/AOG.0000000000003999

Ashoor, G., Syngelaki, A., Poon, L. C., Rezende, J. C., & Nicolaides, K. H. (2013). Fetal fraction in maternal plasma cell-free DNA at 11–13 weeks' gestation: Relation to maternal and fetal characteristics. *Ultrasound in Obstetrics and Gynecology, 41*, 26–32.

Bedei, I., Gloning, K., Luc, J., et al. (2023). Multicenter clinical experience with non-invasive cell-free DNA screening for monosomy X and related X-chromosome variants. *Prenatal Diagnosis*, 1–15. https://doi.org/10.1002/pd.6320

Belling, K., Russo, F., Jensen, A. B., Dalgaard, M. D., Westergaard, D., Rajpert-De Meyts, E., Skakkebæk, N. E., Juul, A., & Brunak, S. (2017). Klinefelter syndrome comorbidities linked to increased X chromosome gene dosage and altered protein interactome activity. *Human Molecular Genetics, 26*(7), 1219–1229.

Benn, P., Borrell, A., Chiu, R. W. K., Cuckle, H., Dugoff, L., Faas, B., Gross, S., Huang, T., Johnson, J., Maymon, R., Norton, M., et al. (2015). Position statement from the Chromosome abnormality screening committee on behalf of the board of the international society for prenatal diagnosis. *Prenatal Diagnosis*. https://doi.org/10.1002/pd.4608

Benn, P., Cuckle, H., & Pergament, E. (2013). Non-invasive prenatal testing for aneuploidy: Current status and future prospects. *Ultrasound in Obstetrics and Gynecology, 42*, 15–33.

Berglund, A., Stochholm, K., & Gravholt, C. H. (2020). The epidemiology of sex chromosome abnormalities. *American Journal of Medical Genetics Part C: Seminars in Medical Genetics, 184*(2), 202–215.

Berglund, A., Viuff, M. H., Skakkebæk, A., Chang, S., Stochholm, K., & Gravholt, C. H. (2019). Changes in the cohort composition of Turner syndrome and severe non-diagnosis of klinefelter, 47, XXX and 47, XYY syndrome: A nationwide cohort study. *Orphanet Journal of Rare Diseases, 14*, 16.

Bianchi, D. W., Platt, L. D., Goldberg, J. D., Abuhamad, A. Z., Sehnert, A. J., Rava, R. P., & MELISSA Study Group. (2012). Genome-wide fetal aneuploidy detection by maternal plasma DNA sequencing. *Obstetrics & Gynecology, 119*(5), 890–901. https://doi.org/10.1097/AOG.0b013e31824fb482

Bianchi, D. W., & Wilkins-Haug, L. (2014). Integration of noninvasive DNA testing for aneuploidy into prenatal care: What has happened since the rubber met the road? *Clinical Chemistry, 60*(1), 78–87. https://doi.org/10.1373/clinchem.2013.202663

Bois, E., Nassar, M., Zenaty, D., Léger, J., Van Den Abbeele, T., Teissier, N., et al. (2018). Otologic disorders in Turner syndrome. *European Annals of Otorhinolaryngology, Head and Neck Diseases, 135*(1), 21–24. https://doi.org/10.1016/j.anorl.2017.08.006

Bojesen, A., Juul, S., & Gravholt, C. H. (2003). Prenatal and postnatal prevalence of Klinefelter syndrome: A national registry study. *Journal of Clinical Endocrinology & Metabolism, 88*(2), 622–626. https://doi.org/10.1210/jc.2002-021491

Bonomi, M., Rochira, V., Pasquali, D., Balercia, G., Jannini, E. A., Ferlin, A., & Klinefelter ItaliaN Group (KING). (2017). Klinefelter syndrome (KS): Genetics, clinical phenotype and hypogonadism. *Journal of Endocrinological Investigation, 40*(2), 123–134.

Bussolaro, S., Raymond, Y. C., Acreman, M. L., Guido, M., Da Silva Costa, F., Rolnik, D. L., & Fantasia, I. (2023). The accuracy of prenatal cell-free DNA screening for sex chromosome abnormalities: A systematic review and meta-analysis. *American Journal of Obstetrics & Gynecology MFM, 5*(3), Article 100844. https://doi.org/10.1016/j.ajogmf.2022.100844

Calanchini, M., Moolla, A., Tomlinson, J. W., Cobbold, J. F., Grossman, A., Fabbri, A., et al. (2018). Liver biochemical abnormalities in Turner syndrome: A comprehensive characterization of an adult population. *Clinical Endocrinology, 89*(5), 667–676. https://doi.org/10.1111/cen.13811

Canick, J. A., Palomaki, G. E., Kloza, E. M., Lambert-Messerlian, G. M., & Haddow, J. E. (2013). The impact of maternal plasma DNA fetal fraction on next generation sequencing tests for common fetal aneuploidies. *Prenatal Diagnosis, 33*, 667–674.

Clarke, A. J. (1997). Prenatal screening. In D. S. Harper, & A. J. Clark (Eds.), *Genetics, Society and clinical practice* (pp. 119–140). Oxford: Bios Scientific Publishers.

Clemente, E. G., Penukonda, S. K., Doan, T., Sullivan, B., & Kanungo, S. (2022). Turner syndrome. *Endocrines, 3*, 240–254.

Comité Consultatif National d'éthique (CCNE). (2013). *Questions éthiques associées au développement des tests génétiques fœtaux sur sang maternel*. Paris: CCNE. http://www.ccne-ethique.fr/fr/publications/questions-ethiques-associees-au-developpement-des-tests-genetiques-foetaux-sur-sang#.VBE6h2NQR8w.

Daly, R. F. (1969). Mental illness and patterns of behavior in 10 XXY males. *The Journal of Nervous and Mental Disease, 149*(4), 318–327.

Dao, T. T., Fields, A., Huynh, A., McGhee, N., & Pellegrini, C. (2023). A meta-narrative review: Efficacy of non-invasive prenatal testing (NIPT) in the detection of sex chromosomal aneuploidy in singleton pregnancy. *Research Methods Poster Session, 2023*(1). https://doi.org/10.52519/00107

De Jong, A., & De Wert, G. M. (2015). Prenatal screening: An ethical agenda for the near future. *Bioethics, 29*, 46–55.

Deng, C., Cheung, S. W., & Liu, H. (2021). Noninvasive prenatal screening for fetal sex chromosome aneuploidies. *Expert Review of Molecular Diagnostics, 21*(4), 405–415. https://doi.org/10.1080/14737159.2021.1911651

Essandoh, M., Castellon-Larios, K., Zuleta-Alarcon, A., Portillo, J. G., & Crestanello, J. A. (2014). Unicuspid aortic stenosis in a patient with Turner syndrome: A case report. *Frontiers in Cardiovascular Medicine, 1*, 14. https://doi.org/10.3389/fcvm.2014.00014

Fiot, E., Zénaty, D., Boizeau, P., et al. (2019). X chromosome gene dosage as a determinant of congenital malformations and of age-related comorbidity risk in patients with Turner syndrome, from childhood to early adulthood. *European Journal of Endocrinology, 180*(6), 397–406.

Grati, F. R. (2014). Chromosomal mosaicism in human feto-placental development: Implications for prenatal diagnosis. *Journal of Clinical Medicine, 3*(3), 809–837. https://doi.org/10.3390/jcm3030809

Gravholt, C. H., Andersen, N. H., Christin-Maitre, S., et al. (2024). Clinical practice guidelines for the care of girls and women with Turner syndrome. *European Journal of Endocrinology, 190*(6), G53–G151.

Gravholt, C. H., Juul, S., Naeraa, R. W., & Hansen, J. (1998). Morbidity in Turner syndrome. *Journal of Clinical Epidemiology, 51*(2), 147–158. https://doi.org/10.1016/s0895-4356(97)00237-0

Gravholt, C. H., Viuff, M., Just, J., et al. (2023). The changing face of Turner syndrome. *Endocrine Reviews, 44*(1), 33–69.

Gross, S. J., Carter, S. M., & Weinstock, D. (September 30, 2022). *The genome*. The ObG Project.

Groth, K. A., Skakkebæk, A., Høst, C., Gravholt, C. H., & Bojesen, A. (2013). Clinical review: Klinefelter syndrome—a clinical update. *Journal of Clinical Endocrinology & Metabolism, 98*(1), 20–30. https://doi.org/10.1210/jc.2012-2382

Guo, N., Cai, M., Lin, M., Xue, H., Huang, H., & Xu, L. (2022). Positive predictive value of noninvasive prenatal testing for sex chromosome abnormalities. *Molecular Biology Reports, 49*(10), 9251–9256. https://doi.org/10.1007/s11033-022-07754-x

Hamza, R. T., Shalaby, M. H., Hamed, L. S., Abdulla, D. B. A., Elfekky, S. M., Sultan, O. M., et al. (2016). Renal anomalies in patients with Turner syndrome: Is scintigraphy superior to ultrasound? *American Journal of Medical Genetics, Part A, 170A*(2), 355–362. https://doi.org/10.1002/ajmg.a.37425

Huang, X., Zheng, J., Chen, M., et al. (2014). Noninvasive prenatal testing of trisomies 21 and 18 by massively parallel sequencing of maternal plasma DNA in twin pregnancies. *Prenatal Diagnosis, 34*, 335–340.

Jacobs, P. A., & Strong, J. A. (1959). A case of human intersexuality having a possible XXY sex-determining mechanism. *Nature, 183*(4657), 302–303.

Kim, H., Park, J. E., Kang, K. M., et al. (2024). Clinical evaluation of noninvasive prenatal testing for sex chromosome aneuploidies in 9,176 Korean pregnant women: A single-center retrospective study. *BMC Pregnancy and Childbirth, 24*, 93. https://doi.org/10.1186/s12884-024-06275-8

Kleczkowska, A., Dmoch, E., Kubien, E., Fryns, J. P., & Van den Berghe, H. (1990). Cytogenetic findings in a consecutive series of 478 patients with Turner syndrome. The leuven experience 1965–1989. *Genetic Counselling, 1*, 227–233.

Lanfranco, F., Kamischke, A., Zitzmann, M., & Nieschlag, E. (2004). Klinefelter's syndrome. *Lancet, 364*(9430), 273–283. https://doi.org/10.1016/S0140-6736(04)16678-6

Lannoo, L., Lenaerts, L., Van Den Bogaert, K., Che, H., Brison, N., Devriendt, K., et al. (2021). Non-invasive prenatal testing suggesting a maternal malignancy: What do we tell the prospective parents in Belgium? *Prenatal Diagnosis, 41*, 1264–1272. https://doi.org/10.1002/pd.6031

Lau, T. K., Cheung, S. W., Lo, P. S., Pursley, A. N., Chan, M. K., Jiang, F., et al. (2014). Non-invasive prenatal testing for fetal chromosomal abnormalities by low-coverage whole-genome sequencing of maternal plasma DNA: Review of 1982 consecutive cases in a single center. *Ultrasound in Obstetrics and Gynecology, 43*, 254–264.

Leggett, V., Jacobs, P., Nation, K., et al. (2010). Neurocognitive outcomes of individuals with a sex chromosome trisomy: XXX, XYY, or XXY: A systematic review. *Developmental Medicine and Child Neurology, 52*(2), 119–129. https://doi.org/10.1111/j.1469-8749.2009.03545.x

Li, Y., Yang, X., Zhang, Y., Lou, H., Wu, M., Liu, F., et al. (2024). The detection efficacy of noninvasive prenatal genetic testing (NIPT) for sex chromosome abnormalities and copy number variation and its differentiation in pregnant women of different ages. *Heliyon, 10*, Article e24155. https://doi.org/10.1016/j.heliyon.2024.e24155

Lindquist, A., Poulton, A., Halliday, J., & Hui, L. (2018). Prenatal diagnostic testing and atypical chromosome abnormalities following combined first-trimester screening: Implications for contingent models of non-invasive prenatal testing. *Ultrasound in Obstetrics and Gynecology, 51*, 487–492. https://doi.org/10.1002/uog.18979

Lo, Y. M., Corbetta, N., Chamberlain, P. F., Rai, V., Sargent, I. L., Redman, C. W., & Wainscoat, J. S. (1997). Presence of fetal DNA in maternal plasma and serum. *Lancet, 350*(9076), 485–487. https://doi.org/10.1016/S0140-6736(97)02174-0

Lu, X., Wang, C., Sun, Y., Tang, J., Tong, K., & Zhu, J. (2021). Noninvasive prenatal testing for assessing foetal sex chromosome aneuploidy: A retrospective study of 45,773 cases. *Molecular Cytogenetics, 14*, 1. https://doi.org/10.1186/s13039-020-00521-2

Mardy, A., & Wapner, R. J. (2016). Confined placental mosaicism and its impact on confirmation of NIPT results. *American Journal of Medical Genetics, Part C, 172*, 118–122. https://doi.org/10.1002/ajmg.c.31505

Mohamed, S. O. O., Elkhidir, I. H. E., Abuzied, A. I. H., Noureddin, A., Ibrahim, G. A. A., Mahmoud, A. A. A., et al. (2018). Prevalence of autoimmune thyroid diseases among the Turner syndrome patients: Meta-analysis of cross-sectional studies. *BMC Research Notes, 11*, 842. https://doi.org/10.1186/s13104-018-3950-0

Norton, M. E., Jacobsson, B., Swamy, G. K., et al. (2015). Cell-free DNA analysis for noninvasive examination of trisomy. *New England Journal of Medicine, 372*, 1589–1597. https://doi.org/10.1056/NEJMoa1407349

Ottesen, A. M., Aksglaede, L., Garn, I., et al. (2010). Increased number of sex chromosomes affects height in a nonlinear fashion: A study of 305 patients with sex chromosome aneuploidy. *American Journal of Medical Genetics, Part A, A*(152A), 1206–1212. https://doi.org/10.1002/ajmg.a.33334

Parens, E., & Asch, A. (2003). Disability rights critique of prenatal genetic testing: Reflections and recommendations. *Mental Retardation and Developmental Disabilities Research Reviews, 9*, 40–47.

Pe'er Dar, K. J., Curnow, S. J., Gross, S. J., et al. (2014). Clinical experience and follow-up with large scale single-nucleotide polymorphism–based noninvasive prenatal aneuploidy testing. *American Journal of Obstetrics and Gynecology, 211*(5), 527.e1–527.e17. https://doi.org/10.1016/j.ajog.2014.08.006

Poulton, A., & Hui, L. (2025). Noninvasive prenatal testing: An overview. *Australian Prescriber, 48*(2), 47–53. https://doi.org/10.18773/austprescr.2025.019

Price, W. H., Clayton, J. F., Collyer, S., De Mey, R., & Wilson, J. (1986). Mortality ratios, life expectancy, and causes of death in patients with Turner's syndrome. *Journal of Epidemiology & Community Health, 40*(2), 97–102. https://doi.org/10.1136/jech.40.2.97

Robinson, A., Bender, B. G., & Linden, M. G. (1992). Prognosis of prenatally diagnosed children with sex chromosome aneuploidy. *American Journal of Medical Genetics, 44*, 365–368. https://doi.org/10.1002/ajmg.1320440319

Ross, J. L., Kushner, H., Kowal, K., Bardsley, M., Davis, S., Reiss, A. L., Tartaglia, N., & Roeltgen, D. (2017). Androgen treatment effects on motor function, cognition, and behavior in boys with Klinefelter syndrome. *Journal of Pediatrics, 185*, 193–199.e4.

Samango-Sprouse, C. A., & Gropman, A. L. (2013). Introduction: Past, present, and future care of individuals with XXY. *American Journal of Medical Genetics, Part C, 163C*(1), 1–2. https://doi.org/10.1002/ajmg.c.31355

Samango-Sprouse, C. A., & Gropman, A. L. (2016). X & Y chromosomal variations: Hormones, brain development, and neurodevelopmental performance. *Colloquium Series on The Developing Brain, 5*(2), Article i122. https://doi.org/10.4199/c00134ed1v01y201506dbr013

Samango-Sprouse, C. A., Gropman, A. L., Sadeghin, T., Mitchell, F. L., et al. (2013). Positive effects of short course androgen therapy on the neurodevelopmental outcome in boys with 47, XXY syndrome at 36 and 72 months of age. *American Journal of Medical Genetics, Part A, 161A*(3), 501–508. https://doi.org/10.1002/ajmg.a.35769

Slutsky, E., Zarudskaya, O., Hopp, S., & Chen, K. (2023). P635: Maternal implications of triple X syndrome detected after abnormal NIPT and inconclusive expanded carrier screening: A case report. *Genetics in Medicine Open, 1*(1).

Snyder, E. A., San Roman, A. K., Piña-Aguilar, R. E., et al. (2021). Genetic counseling for women with 45, X/46, XX mosaicism: Towards more personalized management. *European Journal of Medical Genetics, 64*(3), Article 104140. https://doi.org/10.1016/j.ejmg.2021.104140

Song, Y., Liu, C., Qi, H., et al. (2013). Noninvasive prenatal testing of fetal aneuploidies by massively parallel sequencing in a prospective Chinese population. *Prenatal Diagnosis, 33*(7), 700–706. https://doi.org/10.1002/pd.4160

Tartaglia, N. R., Cordeiro, L., Howell, S., et al. (2010). The spectrum of the behavioral phenotype in boys and adolescents 47, XXY (Klinefelter syndrome). *Pediatric Endocrinology Reviews, 8*(Suppl. 1), 151–159.

Tartaglia, N. R., Howell, S., Sutherland, A., Wilson, R., & Wilson, L. (2010). A review of trisomy X (47, XXX). *Orphanet Journal of Rare Diseases, 5*, 8. https://doi.org/10.1186/1750-1172-5-8

Wang, E., Batey, A., Struble, C., Musci, T., Song, K., & Oliphant, A. (2013). Gestational age and maternal weight effects on fetal cell-free DNA in maternal plasma. *Prenatal Diagnosis, 33*, 662–666.

Wang, Y., Chen, Y., Tian, F., Zhang, J., Song, Z., Wu, Y., et al. (2014). Maternal mosaicism is a significant contributor to discordant sex chromosomal aneuploidies associated with noninvasive prenatal testing. *Clinical Chemistry, 60*(1), 251–259. https://doi.org/10.1373/clinchem.2013.215145

Wigby, K., D'Epagnier, C., Howell, S., Reicks, A., Wilson, R., Cordeiro, L., & Tartaglia, N. (2016). Expanding the phenotype of triple X syndrome: A comparison of prenatal versus postnatal diagnosis. *American Journal of Medical Genetics, Part A, 170*(11), 2870–2881. https://doi.org/10.1002/ajmg.a.37688

Yang, J., Hou, Y., Guo, F., et al. (2021). Non-invasive prenatal detection of fetal sex chromosome abnormalities using the semiconductor sequencing platform (SSP) in Southern China. *Journal of Assisted Reproduction and Genetics, 38*, 727–734. https://doi.org/10.1007/s10815-020-02056-2

Zhang, B., Lu, B. Y., Yu, B., Zheng, F. X., Zhou, Q., Chen, Y. P., & Zhang, X. Q. (2017). Noninvasive prenatal screening for fetal common sex chromosome aneuploidies from maternal blood. *Journal of International Medical Research, 45*(2), 621–630.

CHAPTER

Neurodevelopmental disorders: Angelman Syndrome & Duchenne Muscular Dystrophy

6

Nadir Aman[1,2], Aayushi Gupta[3], Aishwarya Panapil[4], Prashant Utage[1]
[1]*Utage Child Development Centre, Hyderabad, India;* [2]*JSS Medical College, Mysuru, Karnataka, India;* [3]*Redcliffe Lab, Noida, Uttar Pradesh, India;* [4]*Stand Lifescience Pvt. Ltd., Bangalore, India*

1. Introduction

Neurodevelopmental disorders (NDDs) are a diverse group of conditions characterized by impairments in brain growth and development, resulting in cognitive, behavioral, motor, and social dysfunction. Among these, Angelman Syndrome (AS) and Duchenne Muscular Dystrophy (DMD) stand out as genetically distinct but clinically debilitating disorders that profoundly affect affected individuals and their failies. As monogenic conditions with identifiable molecular etiologies, they are exemplary models for integrating genetics into reproductive care and public health strategies, especially in the era of advanced genomic technologies like Non-Invasive Prenatal Screening (NIPS). AS is a rare neurogenetic disorder caused by loss of function of the maternally inherited *UBE3A* gene, typically due to deletions, paternal uniparental disomy (UPD), imprinting defects, or pathogenic variants. Clinically, AS presents with severe developmental delay, speech impairment, seizures, microcephaly, ataxic gait, and unique behavioral traits such as inappropriate laughter and hyperactivity. Despite a consistent phenotype, the underlying genetic heterogeneity makes diagnosis complex, requiring a tiered genetic approach. Givenm its severe neurological impairment and lifelong dependency, early identification through prenatal strategies is crucial for at-risk families. In contrast, DMD is the most common and severe form of X-linked muscular dystrophy, caused by mutations in the *DMD* gene, which encodes dystrophin, a structural protein essential for muscle integrity. DMD predominantly affects males, presenting in early childhood with progressive muscle weakness, delayed motor milestones, calf pseudohypertrophy, and eventual loss of ambulation, followed by respiratory and cardiac complications. Though DMD primarily affects the musculoskeletal system, its classification under NDD is justified due to associated cognitive deficits, behavioral disturbances, and neuropsychiatric comorbidities such as autism spectrum disorder, ADHD, and learning disabilities, attributed to dystrophin isoform deficiency in the brain. With

Non-Invasive Prenatal Screening of Rare Fetal Genetic Diseases. https://doi.org/10.1016/B978-0-443-27660-6.00004-3

advancements in reproductive genomics, NIPS has emerged as a transformative tool that enables the detection of genetic abnormalities in the fetus through the analysis of cell-free fetal DNA (cfDNA)in maternal blood. Traditionally used for detecting aneuploidies, NIPS has now expanded to include single-gene disorders like AS and DMD through targeted and whole-genome approaches, particularly when there is a known familial mutation or increased recurrence risk. This paradigm shift allows for safer, earlier, and more informed reproductive choices without the risks associated with invasive procedures like chorionic villus sampling or amniocentesis. From a public health perspective, integrating NIPS into prenatal care offers a unique opportunity to reduce the burden of rare NDD by enabling timely diagnosis, genetic counseling, and informed decision-making. It also promotes equity in reproductive health by providing access to genetic insights that were previously available only through invasive or postnatal means. This chapter aims to provide a comprehensive overview of the genetic basis, clinical features, diagnostic strategies, and the evolving role of NIPS in the context of AS and DMD. Emphasis is placed on understanding the natural history of each disorder, available diagnostic modalities (including molecular techniques and NIPS), recurrence risk, and ethical considerations in prenatal decision-making. In doing so, it underscores the importance of genetic literacy among clinicians and the integration of precision medicine into routine prenatal care.

2. Introduction of Angelman Syndrome

AS is a rare, complex neurogenetic disorder characterized by developmental delays, intellectual disability, motor impairments, and a variety of distinct behavioral and physical features. First described in 1965 by *Dr. Harry Angelman*, this condition is caused by the loss of function of the *UBE3A* gene, located on chromosome 15 (Williams et al., 2010). AS predominantly affects the nervous system, with individuals typically exhibiting developmental delays that are apparent in infancy, and often develop epilepsy, speech impairment, and a happy demeanor with frequent laughter. AS arises from several genetic mechanisms, including deletions, mutations, or UPD affecting the *UBE3A* gene. In most cases, the syndrome results from a deletion in the maternal copy of this gene, as the paternal copy is silenced in neurons through an epigenetic process known as genomic imprinting. Research continues to explore potential therapeutic interventions, including gene therapy, pharmacological treatments, and behavioral interventions that aim to improve the quality of life for individuals with AS. Understanding the molecular underpinnings and clinical features of AS is critical for early diagnosis and effective management. Advances in genetic research and clinical care offer hope for better outcomes, although challenges remain due to the syndrome's complex presentation and the limited availability of treatment.

3. Clinical features of Angelman syndrome

AS is a neurogenetic disorder with a characteristic set of clinical features that typically emerge during infancy and early childhood. These features can vary in severity from person to person, but they consistently include a combination of developmental, motor, and behavioral abnormalities. Table 6.1.

3.1 Developmental delay

3.1.1 Global developmental delays

The most prominent feature of AS is global developmental delay. Children with AS typically exhibit delays in motor skills, language, and cognitive development from a young age. Motor milestones, such as sitting, crawling, and walking, may be delayed and often appear much later than in typically developing children.

3.1.2 Cognitive impairment

Individuals with AS usually experience severe to profound intellectual disability. Most affected individuals have little or no speech and demonstrate significantly reduced cognitive abilities. Their IQ is generally low, and they are unable to perform tasks that require higher-level cognitive functioning.

Table 6.1 Clinical features of angelman syndrome.

Clinical feature	Approximate frequency (%)
Global developmental delay	~100
Severe intellectual disability	~100
Speech impairment (absent or minimal speech)	>95
Happy demeanor/frequent laughter	80–90
Ataxic gait and movement disorder	70–90
Seizures	80–90
Abnormal EEG patterns	>90
Microcephaly (postnatal onset)	80–85
Hand-flapping movements	70–80
Hyperactivity/short attention span	80–90
Sleep disturbances	60–80
Feeding difficulties in infancy	40–60
Strabismus	40–50
Hypopigmentation (skin, hair, eyes)[a]	20–40
Tongue protrusion/wide mouth	60–80
Fascination with water	60–70

[a] *More common in patients with 15q11-q13 deletions.*

3.2 Motor and coordination abnormalities

3.2.1 Ataxia

Ataxia, or uncoordinated movement, is a common motor abnormality seen in AS. Affected individuals often exhibit difficulty with balance and coordination, which can be apparent even in early childhood. This leads to challenges in walking, running, and other motor activities.

3.2.2 Movement and gait issues

Individuals with AS often have a distinctive, unsteady gait that can resemble a "stiff-legged" walk. Their movements are typically jerky, and they may demonstrate hyperactivity or be prone to frequent falls. Fine motor skills are often delayed, and tasks requiring hand-eye coordination, such as writing or feeding, may be challenging.

3.2.3 Tremors and hypertonia

Some individuals with AS may experience tremors, especially in their hands. Hypertonia (increased muscle tone) is also common, leading to stiffness in the limbs and difficulty with certain movements.

3.3 Seizures

3.3.1 Epilepsy

Seizures are present in the majority of individuals with AS, typically starting between 6 months and 3 years of age. These seizures may take various forms, including generalized tonic-clonic seizures, atypical absence seizures, and myoclonic jerks. Seizure activity can be difficult to control, and individuals often require a combination of medications to manage these symptoms.

3.3.2 Seizure types and frequency

The frequency and type of seizures vary among individuals with AS. While some may experience frequent seizures, others may have them less often. Seizures can be triggered by environmental factors such as fever or stress, but in many cases, the cause remains unexplained.

3.4 Speech impairment

3.4.1 Limited or absent speech

Most individuals with AS are nonverbal or have very limited speech. They may use gestures, signs, or augmentative communication devices to express their needs, but the ability to produce meaningful speech is severely impaired. The inability to speak is one of the most consistent features in AS, contributing to significant challenges in daily communication.

3.4.2 Development of nonverbal communication

Despite limited speech, individuals with AS often develop alternative forms of communication, including facial expressions, body language, and vocalizations, which can be used to express emotions, desires, and reactions.

3.5 Behavioral characteristics

3.5.1 Happy demeanor and frequent laughter

One of the hallmark features of AS is a characteristic "happy" demeanor. Children with AS often exhibit a cheerful disposition, with frequent, inappropriate laughter or smiling. This feature is often described as a "happy puppet" behavior, where the individual appears unusually joyful, even in situations that wouldn't typically elicit such a response.

3.5.2 Hyperactivity and attention issues

Individuals with AS are often hyperactive, demonstrating impulsive behavior, distractibility, and difficulty remaining calm or focused for long periods. Their activity level is typically high, and they may show little interest in calming activities, contributing to challenges in behavioral management.

3.5.3 Stereotypic behaviors

Many individuals with AS engage in repetitive or stereotypic behaviors, such as hand-flapping, clapping, or rocking. These behaviors are considered characteristic of AS and are often seen alongside the other behavioral features.

3.6 Sleep disturbances

3.6.1 Sleep problems

Many individuals with AS experience significant sleep disturbances, including difficulty falling asleep, frequent night waking, and a reduced total sleep time. These issues are often more pronounced in childhood and may persist into adolescence and adulthood.

3.7 Other features

Individuals with AS may exhibit certain facial features, such as a wide mouth, a high forehead, and a prominent chin. These features, although not diagnostic, are often noted by clinicians and can help identify the syndrome. Additionally, some individuals may have heightened sensitivity to sensory stimuli, including loud noises, bright lights, or certain textures. This may contribute to irritability or discomfort in certain environments.

4. Genetic etiology and molecular mechanisms

AS is a classic example of a neurodevelopmental disorder arising from errors in genomic imprinting, an epigenetic phenomenon where gene expression depends on the parent of origin (Table 6.2). The disorder results from a deficiency of function in the maternally inherited *UBE3A* gene located on chromosome 15q11.2–q13, a region known for its imprinting regulation (Fig. 6.1). *UBE3A* encodes the E6-AP ubiquitin-protein ligase, a crucial protein for synaptic function and neurodevelopment, especially in the cerebellum and hippocampus. While *UBE3A* is expressed

Table 6.2 Genetic mechanisms in angelman syndrome.

Genetic mechanism	Frequency (%)	Detection method	NIPS detectable?	Detectable by targeted NIPD	Recurrence risk
Maternal deletion	~70–75	CMA, MLPA, FISH	Possibly	Yes (with high resolution)	<1%
UBE3A mutations	~10	Sanger sequencing, NGS	No	Yes (if familial variant known)	Up to 50% if inherited
Paternal UPD	~1–2	Methylation studies, SNP microarray	No	Experimental (haplotype-based)	<1%
IC defect	~2–3	Methylation + IC deletion analysis	No	Rare (methylation-based)	Up to 50% if IC deletion
Mosaic or unknown	<1	Specialized testing (high-depth NGS)	No	-	Variable

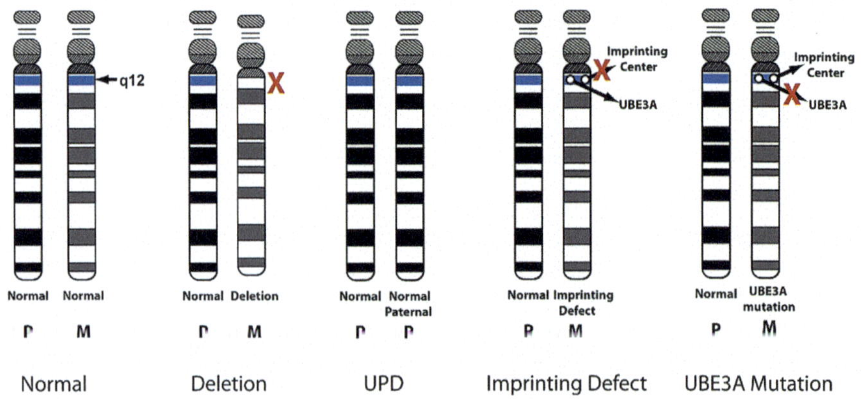

FIGURE 6.1

Genetic mechanisms of 15q11.2–q13 in angelman syndrome.

from both alleles in most tissues, in neurons, only the maternal allele is active, and the paternal allele is silenced via imprinting mechanisms. Loss of function of the maternal *UBE3A* allele leads to the classical clinical phenotype of AS. This loss may occur due to one of four distinct genetic mechanisms, each with implications for diagnosis, recurrence risk, and detection by prenatal screening. Interestingly, Prader–Willi Syndrome and AS both arise from alterations in the same chromosomal region 15q11.2-q13 but differ in the parent of origin of the affected gene. While AS is caused by the absence of function of maternally expressed genes (especially *UBE3A*), PWS results from the loss of paternally expressed genes in the same region. This parent-of-origin effect highlights the critical role of genomic imprinting and makes the 15q11.2–q13 region one of the most well-studied loci in human imprinting disorders. Understanding this shared genetic basis is essential for accurate diagnosis and genetic counseling.

4.1 Maternal 15q11.2–q13 deletion

This is the most common molecular mechanism underlying AS. A de novo deletion on the maternally inherited chromosome 15q11.2–q13 results in the loss of several genes, including *UBE3:00 A*. Maternal 15q11.2–q13 deletions in AS typically span a genomic region of approximately four–six mega bases (Mb). These deletions are further categorized into Class I and Class II types, depending on their breakpoint locations Class I deletions span from BP1 to BP3, while Class II deletions extend from BP2 to BP3 (Fig. 6.2). Clinically, individuals with these deletions often present with

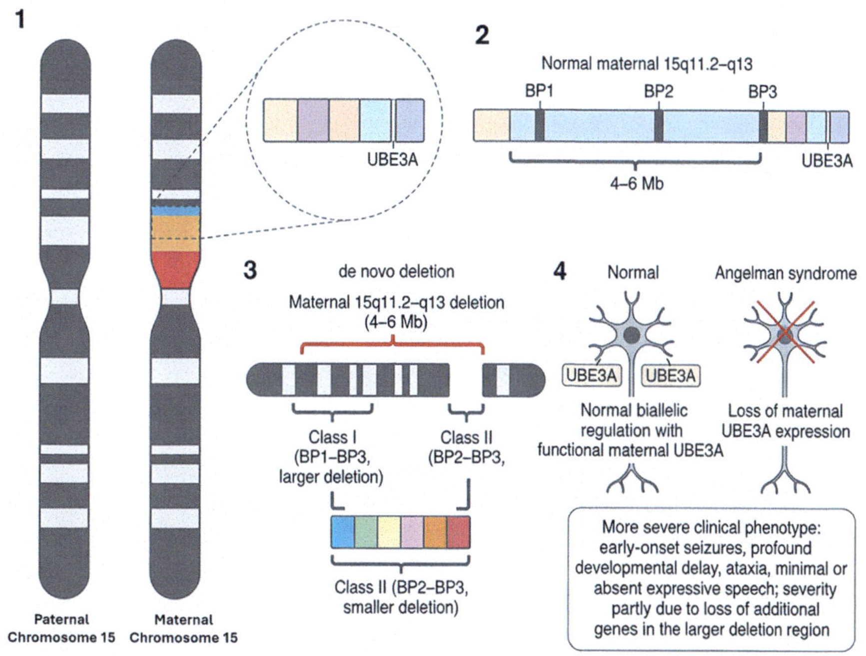

FIGURE 6.2

Genetic mechanism and phenotype of as caused by maternal 15q11.2-q13 deletion.
- ***1.*** *Displays the paternal and maternal chromosome 15 with the critical 15q11.2–q13 region highlighted on the maternal copy.* ***2.*** *A detailed view of the maternal 15q11.2–q13 region, identifying breakpoints BP1, BP2, and BP3, along with the UBE3A gene.* ***3.*** *Illustrates the difference between larger Class I (BP1–BP3) and smaller Class II (BP2–BP3) deletions on the maternal chromosome.* ***4.*** *Shows how the loss of maternal UBE3A expression in neurons results in AS, characterized by seizures, ataxia, and profound developmental delay.*

a more severe phenotype, which includes early-onset seizures, profound developmental delays, ataxia, and minimal to absent expressive language capabilities. The presence of additional genes within the deleted region may contribute to the increased severity observed in these cases.

4.2 *UBE3A* Pathogenic sequence variants

These include point mutations (nonsense, missense, splicing mutations) and small insertions or deletions within the maternal *UBE3A* gene that disrupt protein function (Fig. 6.3). This may present with milder phenotypes compared to deletions. Some individuals may retain partial speech or have fewer seizure episodes.

4.3 Paternal UPD of chromosome 15

In this scenario, both copies of chromosome 15 are inherited from the father, a condition known as paternal UPD. This results in the absence of the maternally derived allele and consequently leads to a lack of functional *UBE3A* expression in neurons, where only the maternal copy is normally active due to imprinting. The underlying mechanism often involves a phenomenon called trisomy rescue, wherein a post-zygotic correction eliminates an extra maternal chromosome in a conceptus initially affected by trisomy 15. Although the overall chromosomal complement becomes numerically normal, the fetus ends up with two paternal copies of chromosome 15 and no maternal contribution, leading to a copy-neutral yet functionally significant loss of *UBE3A* expression (Fig. 6.4).

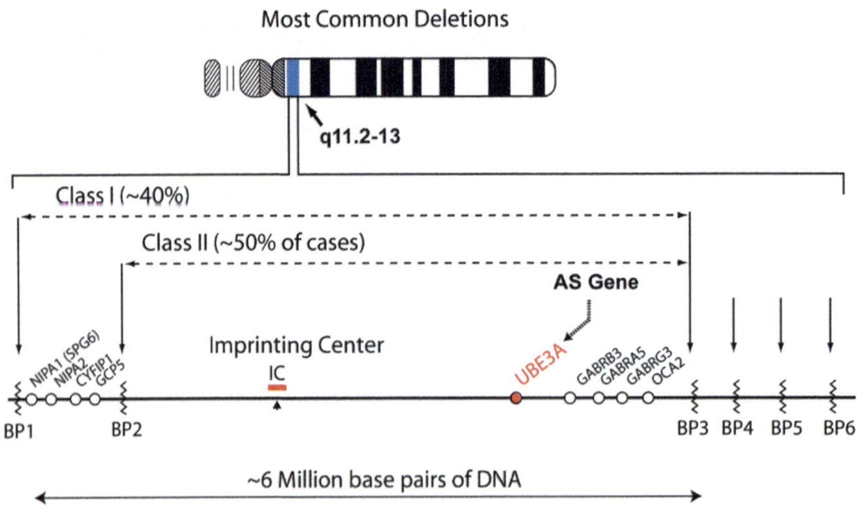

FIGURE 6.3

Genomic map of 15q11.2–q13 deletions in angelman syndrome.

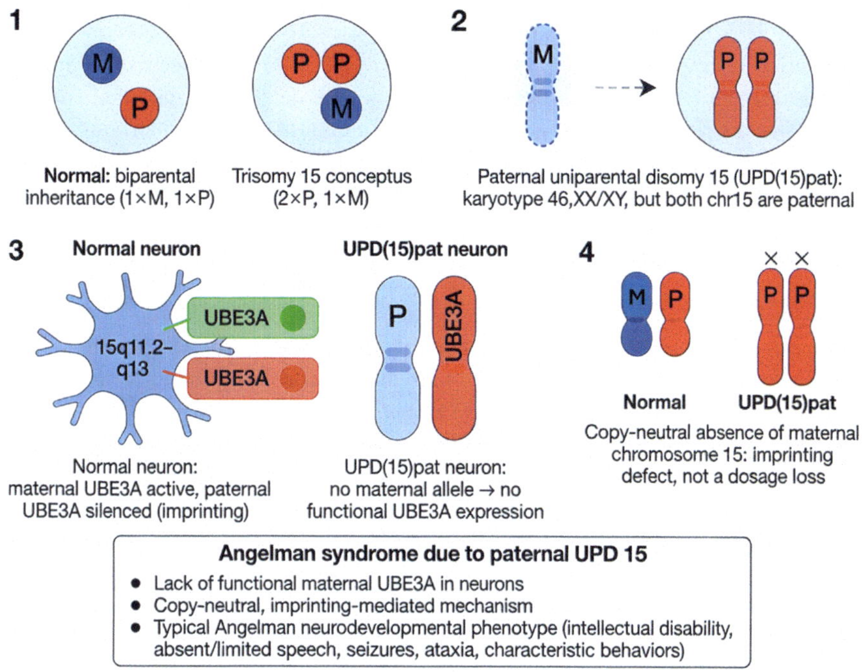

FIGURE 6.4

Genetic mechanism of paternal UPD15.

- 1. Contrasts normal biparental inheritance (one copy from each parent) with a trisomy 15 conceptus that incorrectly contains two paternal copies and one maternal copy. 2. Illustrates "trisomy rescue," where the maternal chromosome is lost, leaving a 46,XX/XY karyotype where both copies of chromosome 15 are paternal UPD. 3. Demonstrates that because the paternal UBE3A alleles are naturally silenced (imprinted), a neuron with only paternal copies has no functional UBE3A expression. 4. Explains that this is a "copy-neutral" defect; the patient has the correct number of chromosomes, but they lack the necessary maternal imprinting "marks".

4.4 Imprinting center defects

Defects in the imprinting control region located upstream of the *SNRPN* gene on chromosome 15 can disrupt the normal methylation and gene expression patterns within the 15q11.2–q13 region, including the *UBE3A* gene. These defects interfere with the maternal-specific methylation process, leading to a functional silencing of *UBE3A* in neurons, even though the gene itself remains structurally intact. Imprinting defects can arise through two primary mechanisms. The first involves epigenetic (non-deletion) defects, which are typically sporadic and result from abnormal methylation patterns without any structural change in the DNA sequence. The second involves microdeletions of the IC, which may be inherited and follow an autosomal dominant pattern of transmission when passed maternally. These inherited deletions pose a significant recurrence risk and highlight the importance of targeted molecular testing and family studies.

4.5 Recurrence risk

If a deletion involving the IC is present in a case of AS, the recurrence risk for future pregnancies can be as high as 50%, especially if the deletion is inherited from a mother who carries the defect in her germline. In contrast, if the underlying cause is an epigenetic change without an associated IC deletion, the recurrence risk is significantly lower - typically less than 1%. This distinction is crucial for genetic counseling and guiding reproductive decisions in affected families.

4.6 Mosaicism

Rare cases of AS are attributed to mosaic imprinting defects or mosaic deletions, wherein only a subset of cells within the individual carries the genetic or epigenetic abnormality. This mosaic pattern results in a partial expression of the disorder, often leading to an atypical or milder clinical phenotype. Affected individuals may exhibit fewer or less severe neurological symptoms, such as improved motor skills, presence of limited speech, or delayed onset of seizures, which can make clinical diagnosis more challenging. Due to the variability in the distribution of affected versus unaffected cells across different tissues, detection of mosaicism requires highly sensitive molecular techniques, such as high-resolution methylation-specific assays or deep sequencing. However, prenatal detection of mosaic cases remains extremely limited and unreliable, as standard prenatal diagnostic tools may fail to capture low-level mosaic changes in chorionic villus or amniotic fluid samples. This underscores the need for cautious interpretation of results and the importance of postnatal follow-up in suspected cases.

5. Diagnostic approaches for Angelman Syndrome

Diagnosing AS involves a combination of clinical evaluation, genetic testing, and the assessment of characteristic clinical features. Due to the variability of symptoms and the overlap with other neurodevelopmental disorders, timely and accurate diagnosis is essential for appropriate management and genetic counseling. Below is a detailed overview of the diagnostic approaches for AS, with references integrated throughout the discussion (Fig. 6.5).

5.1 Clinical evaluation

- **History and Physical Examination**: A thorough clinical history is essential in the diagnosis of AS. Parents and caregivers typically report developmental delays, movement disorders, speech impairment, and characteristic behaviors such as excessive laughter and a happy demeanor. The physical examination may reveal motor abnormalities such as ataxia, tremors, and hypertonia, which are common in individuals with AS.

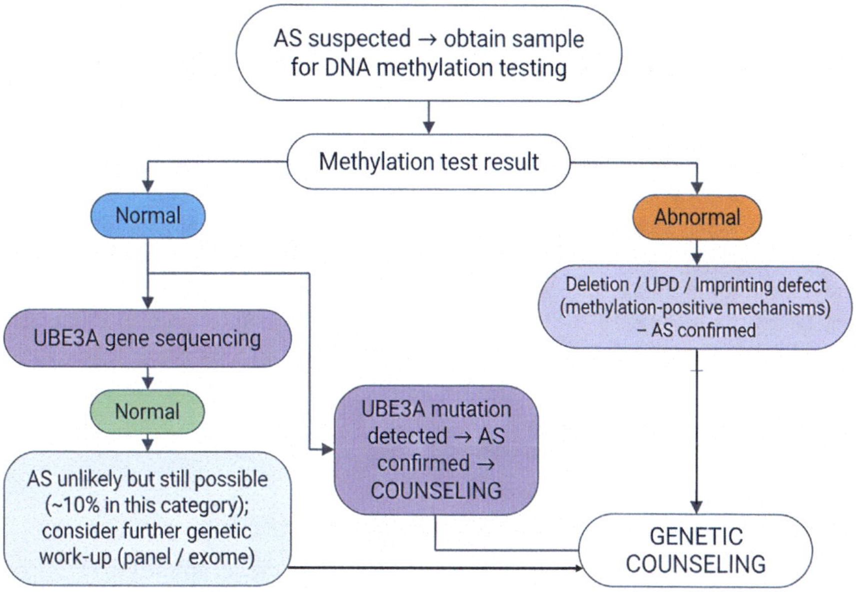

FIGURE 6.5

Diagnostic approaches for as.

- The figure summarizes the stepwise genetic testing methods used to diagnose AS, including clinical evaluation, methylation analysis, deletion analysis, and targeted genetic sequencing, highlighting their roles in confirming diagnosis and guiding genetic counseling.

- **Identifying Key Symptoms**: The presence of a distinctive behavioral phenotype, including frequent laughing, smiling, and a generally happy disposition, is highly suggestive of AS. These features, along with developmental delays and lack of speech, often raise suspicion and prompt further testing.

5.2 Genetic testing

Early and accurate diagnosis of AS is critical for appropriate clinical management, family counseling, and prenatal decision-making in future pregnancies. Given the diverse molecular mechanisms underlying AS including maternal deletions, UPD, imprinting defects, and *UBE3A* gene mutations an integrated diagnostic strategy combining cytogenetic, epigenetic, and molecular techniques is essential.

5.3 Clinical suspicion and phenotypic assessment

Diagnosis of AS often begins with clinical suspicion based on characteristic features such as developmental delay, minimal or absent speech, gait ataxia, epilepsy (especially with atypical patterns like non-convulsive status epilepticus), and a happy, excitable demeanor with frequent laughter. Facial features such as a wide mouth, protruding tongue, and microcephaly may be noted, especially in deletion-positive cases. However, due to clinical overlap with other neurodevelopmental disorders, genetic confirmation is necessary.

5.4 Methylation analysis - first-line test

Methylation-specific testing is the recommended first-tier diagnostic tool for suspected AS cases. Methylation-sensitive PCR (MS-PCR) or methylation-specific multiplex ligation-dependent probe amplification (MS-MLPA) can detect abnormal methylation at the imprinting control region of the SNRPN gene. This method can identify approximately 78%–80% of AS cases, including those due to deletions, paternal UPD, and IC defects. While highly sensitive and specific, methylation testing cannot detect pathogenic variants in *UBE3A*, which account for about 10% of cases. Furthermore, methylation testing may not reliably detect mosaic imprinting defects, requiring more sensitive approaches.

5.5 Chromosomal microarray and deletion detection

CMA is used to detect copy number variation, including the recurrent 15q11.2–q13 deletions, which are present in approximately 70% of AS cases. CMA offers high-resolution coverage, enabling detection of both Class I and Class II deletions. It also helps rule out other genomic disorders that may mimic AS phenotypically. However, it cannot detect UPD or methylation abnormalities, and thus must be complemented with epigenetic testing.

5.6 UPD and SNP microarray

In cases where methylation studies indicate loss of maternal contribution without a detectable deletion, single nucleotide polymorphism microarray is recommended to evaluate for paternal UPD. This technique detects copy-neutral loss of heterozygosity, confirming the presence of two paternal chromosomes 15 without maternal input. While SNP microarrays can identify segmental or complete isodisomy, they do not detect methylation patterns or point mutations.

5.7 *UBE3A* Sequencing – detecting point mutations

When both methylation testing and deletion analysis are negative, sequencing of the *UBE3A* gene using Sanger sequencing or next-generation sequencing is necessary. Approximately 10% of AS cases are caused by pathogenic variants in the *UBE3A* gene, most of which are de novo, although some familial cases have been documented. NGS panels may also include genes associated with differential diagnoses to aid in broader screening.

5.8 Detection of IC defects

If methylation testing is abnormal but no deletion is observed on CMA, targeted analysis of the IC should be performed using MS-MLPA or targeted deletion analysis. Microdeletions of the IC can be familial and require testing of parents to assess recurrence risk. These cases demand meticulous genetic counseling due to their potential autosomal dominant inheritance with maternal transmission.

5.9 Prenatal diagnostic approaches

Prenatal diagnosis is particularly relevant in families with a known mutation or chromosomal rearrangement. For such cases, chorionic villus sampling or amniocentesis can be performed, followed by targeted molecular analysis such as MS-MLPA, CMA, or *UBE3A* sequencing (Table 6.3). While NIPS using cffDNA holds promise for detecting large deletions, its utility for detecting UPD, imprinting defects, or point mutations is currently limited and not recommended as a standalone diagnostic tool. In suspected familial cases, preimplantation genetic testing may also be considered as a reproductive option.

5.10 Mosaicism and diagnostic challenges

Mosaic forms of AS, involving imprinting defects or deletions in only a proportion of cells, can complicate diagnosis. These cases often present with atypical or milder phenotypes, and their detection requires high-sensitivity testing techniques. Standard prenatal and postnatal tests may fail to detect low-level mosaicism, making diagnosis elusive. In such situations, a multidisciplinary evaluation, including clinical genetics and neurology, is critical to guide further testing and management.

6. Comparison with traditional diagnostic methods

AS is a rare neurodevelopmental disorder that requires accurate and timely diagnosis for proper management and care. In the past, traditional diagnostic methods relied primarily on clinical observation and basic genetic testing techniques. However, advancements in genetic analysis and molecular biology have significantly enhanced diagnostic accuracy and efficiency. Below is a comparison between traditional diagnostic methods and current advanced genetic diagnostic approaches for AS.

6.1 Clinical diagnosis versus genetic testing

- **Traditional Methods**: Traditionally, the diagnosis of AS was based mainly on clinical signs and symptoms, such as developmental delay, motor impairments (e.g., ataxia), speech deficits, and behavioral characteristics like frequent laughter and a "happy" demeanor. Although clinical features remain crucial in raising suspicion of AS, these signs alone are insufficient for a definitive

Table 6.3 Comparison of prenata diagnostic techniques in as.

Technique	Target genetic mechanism	Detects as subtypes	Sensitivity/ Specificity	Use in prenatal screening	Limitations
NIPS	Large copy number changes (e.g., deletions)	Maternal 15q11.2–q13 deletion (some platforms)	Moderate for deletions	Emerging for high-risk pregnancies	Cannot detect UPD, imprinting defects, or *UBE3A* variants
CMA	Genome-wide CNVs, including 15q11 2–q13 deletion	Deletions (>100 kb), mosaicism (if >30%)	High	Preferred for structural abnormalities	Cannot detect sequence variants or methylation defects
MLPA	Gene-specific deletions/duplications	15q11.2–q13 deletions, IC defects	High	Used after abnormal ultrasound or NIPS	Limited resolution; sequence variants undetected
MS-PCR/ MS-MLPA	Methylation abnormalities	All AS cases except *UBE3A* mutations	Very high	Diagnostic in suspected AS pregnancies	Cannot distinguish among different molecular causes
Sanger sequencing/ NGS (*UBE3A*gene)	Sequence variants (point mutations indels)	*UBE3A* pathogenic variants (~10% cases)	Very high	Targeted testing when deletion/methylation normal	Does not detect deletions or imprinting defects
SNP microarray	UPD, consanguinity, deletions	Paternal UPD, deletions	High	Used when methylation pattern abnormal	Cannot detect point mutations or methylation status
FISH	Large deletions or rearrangements	15q11.2–q13 deletions	Moderate	Limited use; historical method	Low resolution; does not detect epimutations or UPD

diagnosis due to their overlap with other NDD such as Rett syndrome and Prader–Willi syndrome.

- **Advanced Genetic Testing**: Today, molecular genetic testing plays a pivotal role in the diagnosis of AS. Techniques such as DNA methylation analysis, *UBE3A* gene deletion testing, and UPD analysis have revolutionized the diagnostic process. Methylation testing, for instance, can detect 80%–85% of AS cases caused by maternal chromosome 15 deletions or abnormal methylation patterns. These genetic methods provide an objective and definitive diagnosis, reducing ambiguity in cases where clinical presentation may overlap with other conditions.

6.2 Chromosome analysis versus molecular genetic testing

- **Traditional Methods**: Chromosomal analysis, such as karyotyping, was once a standard technique used to detect chromosomal abnormalities. However, karyotyping often fails to detect smaller chromosomal deletions, which are common in AS. In cases of AS, this method would not identify the typical 15q11-q13 deletion that causes the condition. Karyotyping also provides limited information about epigenetic abnormalities like DNA methylation, which are central to AS pathogenesis.
- **Advanced Molecular Methods**: The use of molecular techniques, such as FISH and quantitative PCR, provides more precise detection of deletions and mutations in the *UBE3A* gene. These methods can identify small deletions in the chromosome 15 region, which would not be detectable with traditional karyotyping. DNA methylation analysis has further improved diagnostic sensitivity by identifying abnormal patterns specific to AS, even in the absence of visible deletion.

6.3 Exclusionary testing versus direct diagnosis

- **Traditional Methods**: Before the advent of precise genetic testing, the diagnosis of AS often involved exclusionary testing. Clinicians would consider other conditions that shared similar features, such as Rett syndrome or Prader–Willi syndrome. Genetic testing for these conditions was performed to rule them out, but it was often time-consuming and could delay a definitive diagnosis.
- **Current Approaches**: Modern genetic techniques allow for direct diagnosis rather than exclusion. Molecular testing (e.g., DNA methylation, UPD, and gene deletion analysis) provides a definitive answer for AS in most cases. In cases where DNA methylation analysis is inconclusive, further testing for deletions, UPD, or mutations in the *UBE3A* gene can confirm the diagnosis, thus eliminating the need for extensive differential diagnosis. This direct approach has sped up the diagnostic process, which is critical for early intervention.

6.4 Clinical observations versus neuroimaging and EEG

1 **Traditional Methods**: In the past, clinicians relied on clinical observations of physical features such as ataxia, tremors, and developmental delays. Imaging techniques like MRI or CT scans were sometimes used to rule out other conditions with similar symptoms, but AS typically does not show specific abnormalities on these scans. Furthermore, EEG was used primarily for evaluating the seizure activity common in AS, but it was not sufficient to confirm a diagnosis.
2 **Modern Diagnostic Approaches**: While neuroimaging (e.g., MRI) and EEG remain useful for assessing the neurological impact of AS, they are not diagnostic tools. However, they can help manage complications like seizures and monitor brain structure. EEG findings in AS are typically abnormal, with generalized discharges observed in the majority of individuals, but these findings are non-specific and can overlap with other disorders. The primary diagnostic tool has shifted to genetic testing, which can definitively confirm the presence of the genetic mutation or abnormality associated with AS.

6.5 Genetic counseling and family planning

- **Traditional Methods**: Previously, families suspected of having a child with AS were provided with limited genetic information based on the clinical diagnosis alone. Families were often unaware of the precise genetic cause of the disorder, which complicated family planning and recurrence risk assessments.
- **Modern Approaches:** With the advent of advanced genetic testing, genetic counseling has become an integral part of the diagnostic process. Families can now receive precise information about the genetic cause of the disorder, whether it is a deletion, mutation, or UPD. This information helps to inform recurrence risk assessments, and genetic testing of parents can identify carriers of genetic abnormalities or mutations, which is vital for family planning.

In comparison to traditional diagnostic methods, modern molecular genetic testing has significantly improved the accuracy, speed, and reliability of diagnosing AS. The ability to detect specific genetic abnormalities like deletions, methylation changes, or UPD using advanced methods such as DNA methylation analysis and FISH provides a definitive diagnosis, often without the need for lengthy clinical observations or exclusionary testing. These advances have led to more timely interventions and better-informed genetic counseling for affected families. As a result, modern diagnostic techniques offer substantial advantages over traditional methods by enabling early and accurate diagnosis, which is crucial for managing the symptoms and improving the quality of life for individuals with AS.

7. Diagnostic yields and clinical implementation of genetic testing

Advancements in genetic testing have significantly improved the diagnostic yield and clinical implementation for individuals suspected of having AS. Genetic tests such as DNA methylation analysis, *UBE3A* gene deletion analysis, and UPD testing have become integral tools for diagnosis. The following is a detailed overview of the diagnostic yields, clinical implementation, and the impact these have on patient care, along with accurate references.

7.1 Diagnostic yield of molecular testing

The diagnostic yield of genetic testing for AS has increased dramatically in recent years. Key diagnostic methods include DNA methylation analysis, *UBE3A* gene deletion testing, UPD testing, and *UBE3A* mutation testing.

- **DNA Methylation Analysis**: DNA methylation analysis is the first-line test for diagnosing AS. This method detects abnormal methylation patterns on the maternal allele of the *UBE3A* gene located on chromosome 15, which is the most common cause of AS. Studies show that methylation testing identifies the genetic cause of AS in approximately 80%–85% of cases. This high diagnostic yield is a significant advancement over traditional diagnostic approaches, which relied primarily on clinical presentation.
- ***UBE3A*Gene Deletion Testing**: In cases where DNA methylation testing yields inconclusive results, *UBE3A* gene deletion testing can be performed. Deletions of the maternal allele of *UBE3A* are responsible for approximately 70% of AS cases. Techniques like fluorescence in situ hybridization (FISH) or quantitative PCR are commonly used for detecting these deletions, offering a diagnostic yield of 70%–80% of AS cases.
- **UPD Testing**: UPD occurs when both copies of chromosome 15 are inherited from the father, leading to the absence of a functional *UBE3A* gene on the maternal chromosome. UPD is a rarer cause of AS and accounts for approximately 3%–5% of cases. Testing for UPD can confirm the diagnosis in these instances.
- ***UBE3A*Mutation Testing**: In rare cases where neither methylation nor deletion analysis provides a diagnosis, sequencing of the *UBE3A* gene can be performed to detect mutations. These mutations are responsible for a small proportion of AS cases. This test has a much lower diagnostic yield compared to the other methods but remains a critical component for certain atypical presentations.

Overall Diagnostic Yield: Combining these tests provides a near-complete diagnostic yield of approximately 90%–95% in individuals with suspected AS. This represents a significant improvement over traditional diagnostic approaches that relied more on clinical judgment and exclusion.

7.2 Clinical implementation and impact on patient care

The introduction of genetic testing into clinical practice has dramatically changed the landscape of diagnosing and managing AS. The benefits of these advances are reflected in both diagnostic accuracy and clinical management.

- **Early Diagnosis**: Genetic testing allows for early diagnosis of AS, often before the appearance of more severe symptoms. Early diagnosis is crucial for starting interventions that can improve developmental outcomes. For example, individuals diagnosed early can benefit from early intervention programs that focus on speech, motor skills, and socialization, which can significantly improve quality of life.
- **Genetic Counseling**: With the availability of genetic testing, genetic counseling has become an essential component of the clinical implementation of AS diagnostics. Parents of children diagnosed with AS can be informed about the genetic causes of the disorder, recurrence risks for future pregnancies, and available prenatal testing options. Genetic testing also enables accurate family planning decisions and helps provide a better understanding of the disorder's inheritance patterns.
- **Differentiation from Other Neurodevelopmental Disorders:** AS shares clinical features with other neurodevelopmental disorders, such as Rett syndrome, Prader–Willi syndrome, and autism. Genetic testing helps to differentiate AS from these disorders, ensuring that patients receive appropriate care and interventions for their specific diagnosis. For instance, Rett syndrome has a distinct progression involving loss of purposeful hand movements, and Prader–Willi syndrome involves obesity and a different genetic defect, which are not present in AS.
- **Seizure Management**: Seizures are common in AS, and their presence can be assessed using EEG. Once diagnosed, patients with AS can be treated with the appropriate anti-epileptic medications based on the type of seizures they experience. Early diagnosis facilitates timely intervention for managing seizures, a crucial aspect of improving quality of life.
- **Pharmacogenetic Insights**: Recent advancements in understanding the molecular basis of AS, including the role of the *UBE3A* gene, have led to potential targeted therapeutic strategies. Studies are underway to investigate whether drugs can activate the silenced paternal *UBE3A* allele, which could potentially improve cognitive and motor functions in individuals with AS. Genetic diagnosis can facilitate participation in clinical trials of such therapies, offering new hope for treatment.

7.3 Challenges and limitations

While genetic testing has significantly improved diagnostic yield, there are still some challenges and limitations associated with the process.

- **Cost and Accessibility**: Genetic testing can be expensive, and its availability may be limited in certain regions or health systems. This may delay diagnosis in resource-constrained environments, and some families may face financial barriers to accessing testing.

- **Genetic Heterogeneity:** Although most cases of AS are caused by deletions or DNA methylation defects, the rare mutations and UPD cases pose challenges in diagnosis, particularly when clinical features are mild or atypical. In these cases, genetic testing may be inconclusive or require more specialized approaches.
- **Psychosocial Impact:** The emotional and psychological burden of receiving a diagnosis of a rare genetic condition like AS can be significant for families. Genetic counseling is essential to help families cope with the diagnosis, understand the condition, and make informed decisions regarding future family planning.

The integration of genetic testing into clinical practice has greatly enhanced the diagnostic yield for AS, with near-complete diagnostic accuracy (90%–95%) achievable through a combination of DNA methylation analysis, *UBE3A* deletion testing, UPD testing, and *UBE3A* mutation analysis. The clinical implementation of these tests has improved early diagnosis, facilitated genetic counseling, and allowed for better clinical management, especially with respect to early intervention, seizure management, and potential participation in targeted therapies. Despite some challenges, the clinical benefits of genetic testing for AS are profound and continue to evolve as new therapeutic strategies are developed.

8. Case studies

8.1 Case 1

A 3-year-old female child, born to non-consanguineous parents, was referred for evaluation of global developmental delay, absent speech, frequent smiling, and ataxic gait. Pregnancy was uneventful, but no prenatal genetic screening was performed.

On examination, she showed microcephaly, hypotonia, and inappropriate laughter. EEG demonstrated a characteristic pattern of large-amplitude slow-spike waves. MRI brain revealed mild cerebral atrophy and delayed myelination. Biochemical investigations including metabolic screening were normal. Molecular testing by methylation-specific PCR detected loss of maternal methylation at 15q11.2-q13, confirming AS due to a maternal deletion. CMA further confirm a 5 Mb deletion in the region. The family received detailed genetic counseling regarding the diagnosis, recurrence risk (approximately 1% due to possible germline mosaicism), and reproductive options. The counselor emphasized the potential benefit of NIPS or invasive prenatal testing in future pregnancies for early detection.

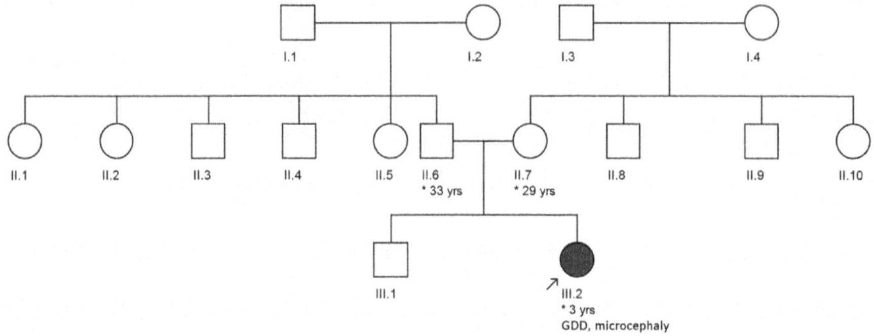

Three generation family pedigree for case 1.

8.2 Case 2

A 2-year-old male child, born after an uneventful pregnancy where NIPS was done and reported negative for common trisomies, presented with severe developmental delay, absent speech, microcephaly, seizures, and poor feeding.

Physical examination revealed hypotonia and dysmorphic facial features including a wide mouth and prominent chin. EEG showed a typical epileptiform pattern. MRI brain demonstrated delayed myelination and mild cerebellar hypoplasia. Methylation-specific MLPA and chromosomal microarray revealed paternal UPD of chromosome 15, consistent with AS. Parents were counseled on the genetic etiology, the limitations of standard NIPS (which typically do not detect UPD or imprinting defects), and the low recurrence risk. Genetic counseling focused on prognosis, management, and the option for prenatal diagnosis in future pregnancies using targeted testing.

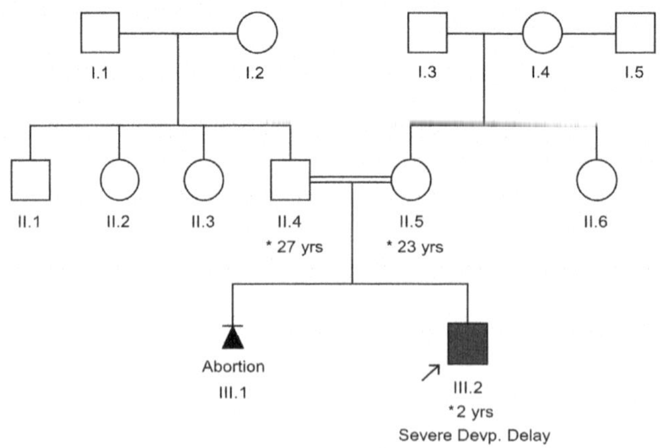

Three generation family pedigree for case 2.

9. Introduction of Duchenne Muscular Dystrophy

DMD is a severe X-linked recessive neuromuscular disorder caused by pathogenic variants in the *DMD* gene, one of the largest known human genes, located at Xp21.2. This gene encodes dystrophin, a critical cytoskeletal protein that plays an indispensable role in preserving the structural and functional integrity of muscle cells. Dystrophin is an essential component of the dystrophin-glycoprotein complex, which serves as a structural and functional bridge between the intracellular cytoskeleton and the extracellular matrix. This connection is vital for maintaining muscle membrane stability, particularly during the mechanical stress associated with muscle contraction and relaxation. In the absence of functional dystrophin, muscle fibers become structurally compromised, rendering them highly susceptible to mechanical damage (Fig. 6.6). This vulnerability initiates a cascade of pathological events, including repeated cycles of myofiber degeneration and regeneration, chronic inflammation, and fibrosis, ultimately leading to progressive muscle wasting - a hallmark of DMD. *A conceptual analogy that elucidates the role of dystrophin in muscle cells is that of a suspension bridge. In this model, the roadway symbolizes the muscle fiber, while the suspension cables represent dystrophin. These cables function to anchor the roadway to the supporting towers - analogous to the extracellular matrix - thereby distributing mechanical stress and preventing structural collapse. In the absence of these stabilizing cables, the bridge would be unable to withstand external forces and would ultimately fail under strain.* Similarly, in individuals with DMD,

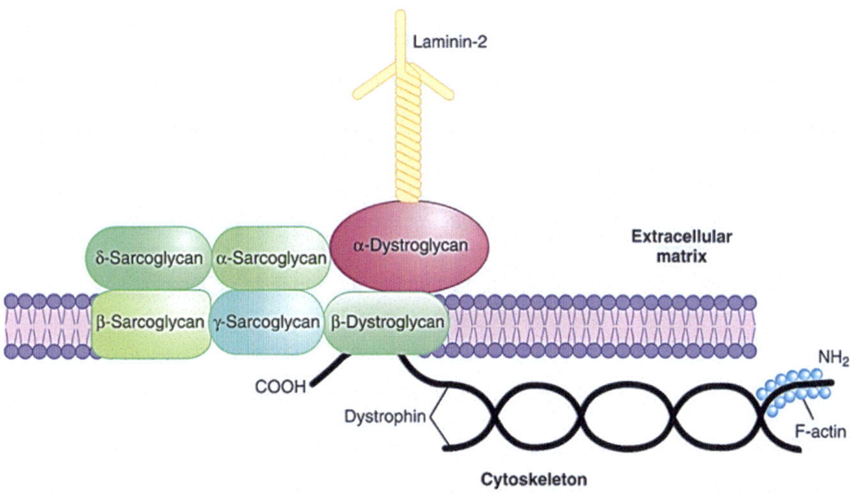

FIGURE 6.6

The amino terminus of dystrophin binds to F-actin in the cytoskeleton, while the carboxyl terminus connects to the dystroglycan-sarcoglycan complex, thereby linking the muscle membrane to the extracellular matrix via laminin.

the absence of dystrophin precipitates a gradual and relentless deterioration of muscle integrity, leading to the characteristic phenotype of the disorder.

DMD presents a significant challenge in genetic counseling for pediatricians, neurologists, and clinical geneticists. Given its X-linked recessive pattern of inheritance, a substantial number of female relatives may be at risk of being carriers, even in families with only a single affected individual. It is imperative that these individuals receive precise and comprehensive genetic counseling, ensuring the accurate interpretation and application of all available diagnostic information in determining carrier status. The advent of molecular genetic techniques has substantially mitigated past challenges associated with insufficient or, more concerningly, erroneous risk assessments. At present, there are very few families for whom conclusive genetic information cannot be provided following appropriate molecular analyses. However, in healthcare settings where molecular diagnostic tools remain inaccessible, traditional methods of carrier risk estimation may still retain relevance. The complete abandonment of these conventional approaches would be inadvisable, as their prudent application remains essential in preventing potentially detrimental consequences for affected families. Significant advancements have been made in the understanding and management of this condition in recent years, establishing it as a model for the study of other genetic disorders. Notably, *DMD* is the largest known human gene with 79 exons, encompassing more than two million base pairs, including non-coding regions. However, despite its extensive size, it does not encode the largest human protein, a distinction that belongs to another muscle protein (Fig. 6.7).

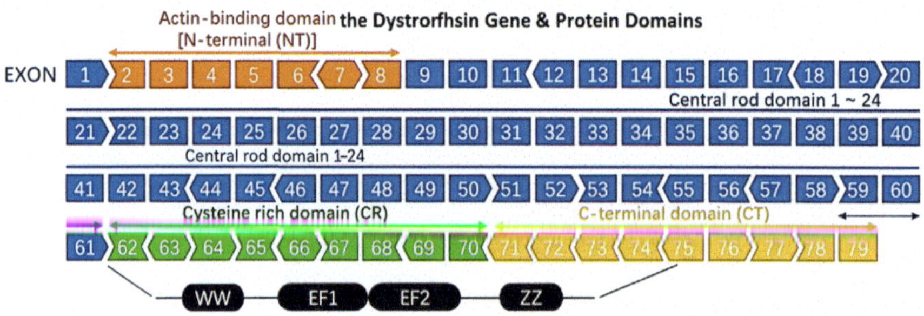

FIGURE 6.7

Structure of the dystrophin gene, which contains 79 exons. *N-terminal domain (NT): exon 2–8; Central rod domain: exon 9–61; Cysteine-rich domain (CR): exon 64–70; C-terminal domain (CT): exon 71–79. The arrow shape of the adjacent exons shows open reading frame (ORF) compatibility. The CR and CT domains comprise the WW domain, EF hand and ZZ domains.*

- From Chengmai sun.

10. Therapeutic strategies for Duchenne Muscular Dystrophy: an update

The considerable length and structural complexity of the gene contribute to a relatively high frequency of recombination events, which may result in DNA markers within the gene exhibiting recombination with a family's specific pathogenic variant. Nevertheless, the majority of cases approximately two-thirds - are attributable to large deletions affecting one or more exons, while a smaller proportion arise from exonic duplications. Consequently, methodologies that assess exon copy number variations enable the identification of the underlying genetic defect in nearly 70% of affected individuals and facilitate the detection of female carriers within the family. Furthermore, sequencing techniques have proven highly effective in identifying pathogenic variants in nearly all remaining cases, most commonly revealing nonsense mutations. As a result, linkage analyses are now seldom required, except in specific contexts such as single-cell testing for preimplantation genetic diagnosis. Management of DMD has evolved considerably over the past two decades, with therapeutic approaches shifting from supportive care alone toward mutation-specific and gene-based interventions. Although DMD remains a progressive disorder, advances in respiratory care, cardiac management, corticosteroid therapy, exon-skipping technologies, and gene therapy have significantly improved survival and quality of life in affected individuals. Corticosteroids continue to represent the cornerstone of disease management. Prednisone and deflazacort have demonstrated benefits in prolonging ambulation, improving pulmonary function, delaying scoliosis, and slowing cardiomyopathy progression. Long-term steroid therapy, however, requires careful monitoring for adverse effects including weight gain, osteoporosis, behavioral changes, hypertension, and growth suppression. Recent years have witnessed the emergence of exon-skipping therapies targeting specific mutations within the *DMD* gene. These therapies utilize antisense oligonucleotides to restore the reading frame during mRNA processing, allowing production of a shorter but partially functional dystrophin protein. Eteplirsen, designed for exon 51 skipping, was among the first approved therapies for eligible patients. Subsequently, golodirsen and viltolarsen targeting exon 53, and casimersen targeting exon 45, expanded treatment options for mutation-specific subgroups. While these therapies do not constitute a cure, they represent an important step toward precision medicine in dystrophinopathies. Nonsense mutation readthrough therapy has also emerged as a targeted therapeutic strategy. Ataluren has shown potential benefit in selected individuals with nonsense mutations by promoting ribosomal readthrough of premature stop codons, thereby enabling dystrophin synthesis. Clinical responses remain variable, and continued long-term evaluation is necessary to better establish therapeutic efficacy. Gene replacement approaches have generated considerable interest in DMD research. Microdystrophin gene therapy utilizes adeno-associated viral vectors to deliver shortened dystrophin constructs capable of restoring partial muscle function. Early clinical trials have demonstrated encouraging results, including increased microdystrophin expression and stabilization of motor function in some patients. Nevertheless, challenges such as immune responses, vector delivery limitations, and long-term durability remain important considerations. Genome editing technologies, particularly CRISPR-Cas9-based approaches, are being investigated as potential future therapeutic options. Experimental studies suggest that targeted correction of pathogenic variants may restore dystrophin expression at the genomic level. Although these approaches remain largely investigational, they highlight the rapidly expanding therapeutic landscape of DMD. In addition to molecular therapies, multidisciplinary supportive management remains essential. Cardiac surveillance with echocardiography and cardiac MRI, respiratory monitoring, physiotherapy, nutritional support, orthopedic management, and psychosocial care substantially influence long-term outcomes. The integration of genetic diagnosis with personalized therapeutic strategies has transformed DMD into a model disorder for translational genomic medicine (Aartsma-Rus & Krieg, 2017; Birnkrant et al., 2018a, 2018b; Mendell et al., 2013; Min et al., 2019).

11. Clinical manifestations and disease progression

DMD typically manifests between 2 and 5 years of age, with early signs including delayed motor milestones, difficulty in running, frequent falls, and a waddling gait due to proximal muscle weakness (Table 6.4). Gower's sign, where children use their

Table 6.4 Stages of disease progression in DMD.

Stage	Age range	Clinical features
Early childhood	2–5 years	- Delayed walking, frequent falls, difficulty running. - Waddling gait, difficulty climbing stairs. - Gower's sign. - Calf muscle pseudohypertrophy.
Childhood progression	6–12 years	- Progressive proximal muscle weakness. - Difficulty raising arms and performing fine motor tasks. - Joint contractures (Achilles tendon, hip flexors). - Scoliosis development. - Loss of independent ambulation (by 9–12 years).
Adolescence	13–20 years	- Respiratory muscle weakness, leading to hypoventilation. - Recurrent respiratory infections, pneumonia risk. - Dilated cardiomyopathy, arrhythmias. - Initiation of non-invasive ventilation (BiPAP). - Mild cognitive impairment, risk of ADHD/ASD.
Adulthood & late stage	>20 years	- Severe respiratory failure requiring ventilatory support. - Advanced cardiomyopathy and heart failure. - Significant upper limb functional loss. - Increased mortality due to cardiac or respiratory complications.

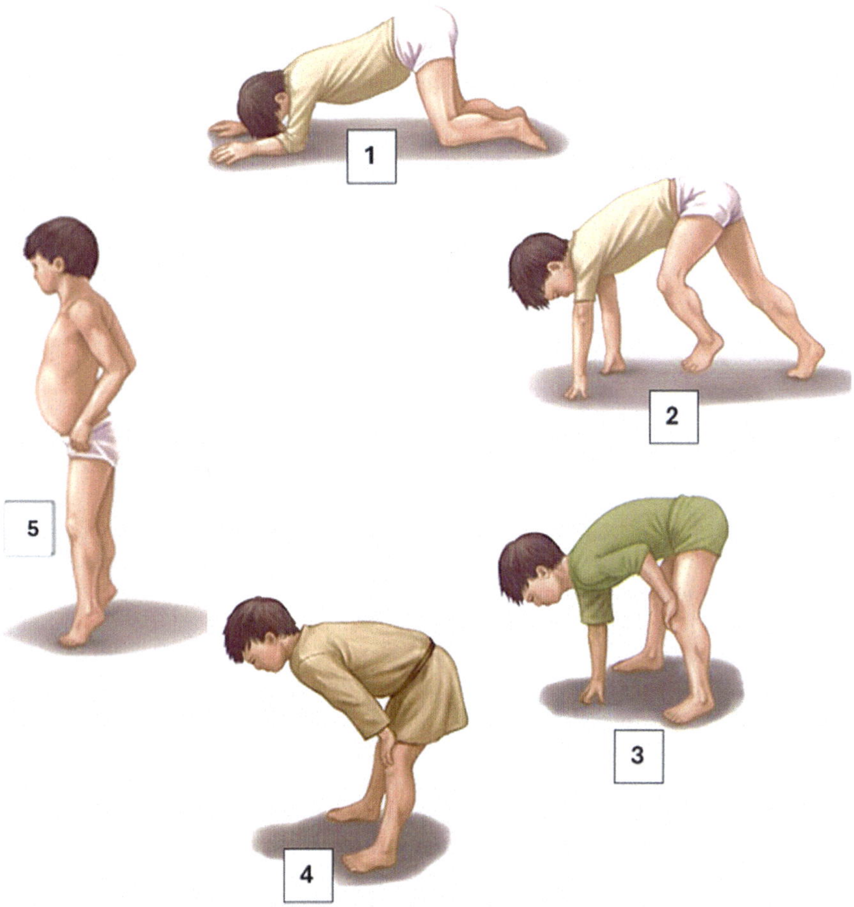

FIGURE 6.8

Illustration of the typical gowers' sign seen in DMD.

hands to "walk up" their thighs when rising, becomes evident as lower limb weakness progresses (Fig. 6.8). Despite muscle atrophy, pseudohypertrophy of the calf muscles is commonly observed due to fat and fibrotic tissue accumulation. Between 6 and 12 years, weakness extends to the shoulder girdle and upper limbs, impairing functional tasks such as raising arms, gripping objects, and writing. Chronic muscle fibrosis leads to joint contractures, especially in the Achilles tendons and hip flexors, contributing to restricted mobility. Scoliosis may develop, further impacting posture and respiratory function. By 9–12 years, most individuals experience loss of independent ambulation, necessitating wheelchair dependence.

During adolescence and early adulthood, respiratory insufficiency emerges due to diaphragmatic and intercostal muscle weakness, increasing the risk of pneumonia and respiratory failure, often requiring non-invasive ventilation (BiPAP). Cardiac involvement becomes prominent, with dilated cardiomyopathy and arrhythmias significantly raising the risk of heart failure. While DMD primarily affects skeletal muscles, cognitive and neurodevelopmental impairments are reported in some individuals, particularly in verbal learning, memory, and processing speed. Additionally, there is an increased prevalence of ADHD and Autism Spectrum Disorder, suggesting a role of dystrophin in brain function. By the third or fourth decade, respiratory and cardiac complications become the leading causes of mortality. While advancements in respiratory and cardiac care have extended life expectancy, the disease remains progressive and ultimately fatal.

12. Diagnostic approaches in DMD

The diagnosis of DMD has undergone significant evolution, transitioning from clinical observations to sophisticated molecular and imaging techniques. Traditionally, physicians relied on characteristic clinical signs such as progressive muscle weakness, difficulty in standing (Gower's sign), and pseudohypertrophy of the calf muscles manifestations that, while indicative, were not exclusive to DMD, thereby complicating differential diagnosis. Early biochemical markers, particularly serum creatine kinase (CK) levels, served as an indirect indicator of muscle damage; however, while elevated CK levels suggested myopathic processes, they lacked specificity for DMD. A major advancement in DMD diagnostics emerged with muscle biopsy and histopathological evaluation, enabling direct assessment of dystrophic muscle tissue. By the late 1970s and early 1980s, the introduction of immunohistochemistry and Western blotting facilitated the detection of dystrophin deficiency, significantly improving diagnostic accuracy. Despite these advancements, muscle biopsy remained invasive and, in some cases, inconclusive due to variability in dystrophin expression. A transformative breakthrough in DMD diagnosis occurred in 1986, when *Louis M. Kunkel* and colleagues identified the *DMD* gene through chromosomal mapping and positional cloning, paving the way for genetic testing. The subsequent development of Multiplex PCR by Chamberlain et al. (1988) enabled the detection of large deletions, though it remained limited in identifying point mutations or small-scale variations. To address these limitations, MLPA was introduced by Schouten et al. (2002), enhancing the detection of both deletions and duplications across all 79 exons of the dystrophin gene. Despite its efficacy, MLPA remained incapable of detecting single-nucleotide variations or deep intronic mutations, necessitating further advancements.

The advent of Next-Generation Sequencing revolutionized DMD diagnostics by allowing for comprehensive whole-gene analysis at an unprecedented resolution. Whole-Exome Sequencing, developed in the late 2000s, further refined mutation detection, facilitating the identification of small-scale and novel genetic alterations that were previously undetectable. These sequencing technologies minimized the need for invasive muscle biopsies, significantly improving early and definitive diagnosis. Beyond genetic analysis, the search for non-invasive biomarkers has gained traction in recent years. MicroRNAs (miRNAs), emerged as potential biomarkers for DMD due to their role in muscle development and pathology. Studies in the 2010s identified disease-associated miRNAs, such as miR-206, miR-1, and miR-133a/b, facilitating the development of serum-based biomarker panels for early detection and disease monitoring. Additionally, exosomal dystrophin, detected in extracellular vesicles, presented novel opportunities for tracking disease progression through blood-based assays, further reducing reliance on invasive procedures. Parallel to these molecular advancements, magnetic resonance imaging and magnetic resonance spectroscopy emerged as non-invasive tools for assessing muscle pathology. Initially pioneered in neuromuscular disease research in the late 1990s and early 2000s, these imaging modalities now play a critical role in monitoring muscle degeneration, fibrosis, and therapeutic responses in both clinical and research settings. The integration of genetic, molecular, and imaging-based diagnostic approaches has transformed the landscape of DMD diagnostics, enabling early detection, precise genetic characterization, and comprehensive disease monitoring, ultimately improving clinical outcomes and therapeutic interventions.

12.1 Muscle histopathology in DMD

The histopathological features of DMD reflect progressive muscle degeneration and defective regeneration. Affected muscle tissue exhibits fiber size variation, central nucleation, and increased endomysial connective tissue, indicating ongoing cycles of muscle necrosis and repair. As the disease advances, normal muscle fibers are progressively replaced by fat and fibrotic tissue, leading to pseudohypertrophy and loss of contractile function (Fig. 6.9). Inflammatory infiltrates are commonly observed, contributing to further muscle damage. These histological changes are key indicators of the irreversible muscle wasting and weakness characteristic of DMD.

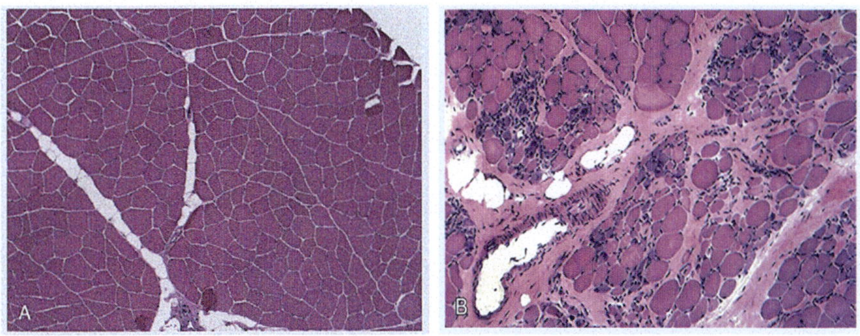

FIGURE 6.9

Two panels showing transverse section of the gastrocnemius muscle from (A) a normal boy and (B) a boy with duchenne muscular dystrophy (DMD). In DMD, normal muscle fibers are progressively replaced by adipose tissue and connective tissue.

- Copyright: Jerdey et al., Medical Genetics.

13. NIPS for DMD

The evolution of DMD diagnostics has significantly advanced in recent decades; however, most conventional diagnostic techniques remain postnatal, thereby limiting opportunities for early intervention, reproductive decision-making, and genetic counseling. Given the X-linked recessive inheritance of DMD, carrier mothers have a 50% probability of transmitting the pathogenic mutation to their offspring, underscoring the necessity for accurate prenatal screening.

In response to these challenges, NIPS has emerged as a promising approach for the early detection of DMD. This method leverages cffDNA short DNA fragments originating from the placenta that circulate in maternal blood to screen for genetic abnormalities without the risks associated with invasive procedures (Chitty et al., 2015). Traditionally, prenatal diagnosis of DMD has relied on chorionic villus sampling or amniocentesis, both of which necessitate the direct extraction of fetal genetic material. CVS, performed between 10 and 13 weeks of gestation, involves sampling placental trophoblastic cells via a transcervical or transabdominal approach. Amniocentesis, typically conducted between 15 and 20 weeks, entails the withdrawal of amniotic fluid containing fetal cells through a transabdominal needle aspiration. While highly accurate, these procedures are inherently invasive and carry a small yet notable risk of miscarriage, infection, or amniotic fluid leakage. NIPS presents a transformative alternative by enabling fetal DNA analysis as early as 7–10 weeks of gestation, thereby significantly reducing procedural risks and enhancing accessibility. The technique has been widely adopted for detecting common chromosomal aneuploidies, including trisomy 21 (Down syndrome), trisomy 18 (Edwards syndrome), and trisomy 13 (Patau syndrome). However, its application in single-gene disorders such as DMD remains an area of active research, primarily due to technical complexities associated with the DMD gene's extensive size and mutational diversity. Spanning 2.2 million base pairs across 79 exons, the DMD gene presents significant analytical challenges, particularly in distinguishing pathogenic fetal variants from maternal DNA in mixed plasma samples. Recent technological advancements, including digital PCR (dPCR), droplet digital PCR (ddPCR), and Next-Generation Sequencing (NGS), have enhanced the precision of NIPS for detecting deletions, duplications, and point mutations within the DMD gene (Chen et al., 2020). A landmark study demonstrated the feasibility of cffDNA-based prenatal screening for DMD. In this study, a mother at risk of transmitting DMD underwent NIPS, which successfully identified a pathogenic mutation, allowing for early prenatal diagnosis and informed reproductive decision-making. Similar findings have been reported in subsequent investigations, highlighting the growing potential of NIPS in reducing the dependence on invasive prenatal testing.

Despite these advancements, several challenges impede the widespread clinical adoption of NIPS for DMD. Low fetal DNA fraction in early pregnancy, potential maternal DNA contamination, and incomplete mutation detection pose significant technical barriers, necessitating further refinement of analytical protocols. Additionally, NIPS functions as a screening tool rather than a definitive diagnostic test, necessitating confirmation via CVS, amniocentesis, or postnatal genetic testing. Future developments, particularly the integration of artificial intelligence (AI)-driven variant analysis and machine learning algorithms, are expected to enhance accuracy and efficiency, paving the way for routine non-invasive prenatal detection of DMD.

As large-scale clinical studies continue to validate its efficacy, NIPS holds the potential to revolutionize prenatal diagnostics by offering at-risk families an early, safe, and accessible genetic screening tool, thereby facilitating proactive disease management and reproductive autonomy.

14. Comparison with traditional diagnostic methods of DMD

The diagnosis of DMD has evolved significantly with advancements in genetic testing. Traditionally, diagnosis relied on clinical evaluation, serum biomarkers, and muscle biopsy, whereas modern approaches emphasize genetic analysis for definitive diagnosis (Table 6.5).

14.1 Traditional diagnostic methods

Clinical assessment is crucial in the diagnosis of DMD, often based on observations of motor delays, progressive muscle weakness, a waddling gait, and the presence of Gower's sign. Progressive loss of ambulation is a key diagnostic clue that raises clinical suspicion. Serum CK levels are typically markedly elevated, reflecting muscle damage; although elevated CK is useful for initial screening, it is not specific to DMD. EMG and nerve conduction studies can help differentiate between neuromuscular and neurogenic disorders. In DMD, EMG usually reveals a myopathic pattern, though this finding is non-specific. Muscle biopsy provides further diagnostic information, with histopathological analysis showing muscle fiber degeneration, fibrosis, and fatty infiltration. Immunohistochemistry can confirm dystrophin deficiency, while Western blot analysis provides quantitative assessment of dystrophin levels, although both procedures are invasive.

Table 6.5 Traditional versus Modern Diagnostic Methods in DMD.

Method	Traditional approach	Modern approach
Primary basis of diagnosis	Clinical and biochemical markers	Genetic testing (mutation analysis)
Specificity & accuracy	Low (CK elevation not DMD-specific)	High (identifies causative mutation)
Invasiveness	Muscle biopsy required	Non-invasive (blood/saliva for genetic testing)
Mutation detection	Not possible	Detects deletions, duplications, and point mutations
Carrier & prenatal testing	Limited utility	Enables carrier screening and prenatal diagnosis
Time to diagnosis	Delayed (requires muscle biopsy confirmation)	Faster (direct genetic confirmation)
Suitability for early diagnosis	Difficult (clinical signs emerge later)	Possible via newborn screening

14.2 Modern genetic diagnostic approaches

MLPA is the first-line molecular test for DMD, as it effectively detects large deletions and duplications in the DMD gene with high sensitivity. For cases where no large rearrangements are identified, Next-Generation Sequencing NGS is utilized to detect point mutations, small deletions or insertions, and deep intronic variants, providing a comprehensive genetic diagnosis. Quantitative PCR (qPCR) is valuable for targeted mutation detection, while Sanger sequencing serves as a confirmatory method for small sequence variants. Emerging newborn screening programs now focus on early detection through measurement of serum CK levels combined with genetic markers, aiming to initiate earlier interventions. Additionally, carrier testing in at-risk females plays a crucial role in genetic counseling and reproductive planning.

15. Diagnostic yields and clinical implementation of DMD

The diagnostic yield of DMD testing refers to the probability of accurately detecting pathogenic variants in the *DMD* gene, thereby confirming the diagnosis. Over time, advancements in genetic testing methodologies have revolutionized diagnostic accuracy, allowing for earlier detection, precise mutation characterization, and improved clinical management strategies. These improvements have substantial implications for early intervention, therapeutic decision-making, genetic counseling, and reproductive planning. Table 6.6.

Table 6.6 Diagnostic yields of DMD testing methods.

Diagnostic method	Sensitivity	Specificity	Key benefits	Limitations
Serum CK testing	~100%	Low	Simple, cost-effective screening	Non-specific for DMD
Muscle biopsy (IHC/ Western blot)	~80%	High	Confirms dystrophin deficiency	Invasive, replaced by genetic tests
MLPA	~65–70%	High	Detects large deletions/ duplications	Misses small mutations
NGS (whole exome/ Genome sequencing)	~98%	High	Identifies all mutation types	Higher cost, requires expertise, cannot detect exonic deletions/ duplications
Sanger Sequencing/ qPCR	-	High	Confirms point mutations	Not first-line testing

16. Case studies

16.1 Case 1

The child (6.3 years, Male) born to parents in a non-consanguineous marriage presents with asymptomatic transaminitis in the last 7 months. Laboratory investigations revealed elevated SGOT at 247 U/L, SGPT at 272 U/L, and bilirubin levels. Abdomen USG showed coarse echotexture and increased hepatic echogenicity. He was suspected to be affected with metabolic liver disease and was advised to undergo whole exome sequencing to establish the genetic basis. Genetic test report: The genetic test report revealed a heterozygous likely pathogenic variant in the *FOCAD* gene (c.4132+2C > A, Autosomal Recessive) associated with Severe Congenital Liver Disease [OMIM#619991]. An additional clinically significant hemizygous pathogenic variant was detected in the *DMD* gene (c.531–2A > C, X-linked Recessive) associated with DMD [OMIM#310200]. He was sent to the Genetic Clinic for further assessment and genetic counseling. On examination, he was positive for Gowers' sign, Valley sign and muscle pseudohypertrophy. A serum creatine kinase level of 1892.34 U/L further supported the diagnosis of DMD. His family history also revealed affected members with similar phenotype associated with DMD.

This case is unique as the diagnosis of DMD was incidental, arising initially from genetic testing performed for suspected liver dysfunction rather than neuromuscular symptoms. This underscores the importance of genetic testing in uncovering unexpected yet clinically significant diagnoses.

16.2 Case 2

The child (10 years, Male) born to parents in a non-consanguineous marriage presents with progressive muscle weakness, difficulty standing and walking, motor regression, and frequent falls. His NCS/EMG study suspected Primary muscle disease affecting proximal, distal and semi-distal muscles, serum creatine kinase levels were found to be 1563.2 U/L and a muscle biopsy that followed confirmed DMD as the final diagnosis. No significant family history was noted in this case. During his visit to the genetics clinic for further evaluation, a physical examination revealed mild pseudohypertrophy and a positive Gower's sign. His mother also complained of muscle spasms and pseudohypertrophy. As a result, MLPA testing was recommended to confirm the diagnosis of DMD genetically. Additionally, carrier screening for his mother and sister was advised to assess their genetic status. Other management strategies were also discussed.

This case is unique as it highlights the importance of comprehensive family evaluation in DMD, given the mother's symptoms suggestive of a manifesting carrier state, which is uncommon but clinically significant. Additionally, it underscores how early genetic testing, such as MLPA, could have prevented the need for an invasive muscle biopsy, emphasizing the role of non-invasive molecular diagnostics in confirming dystrophinopathies.

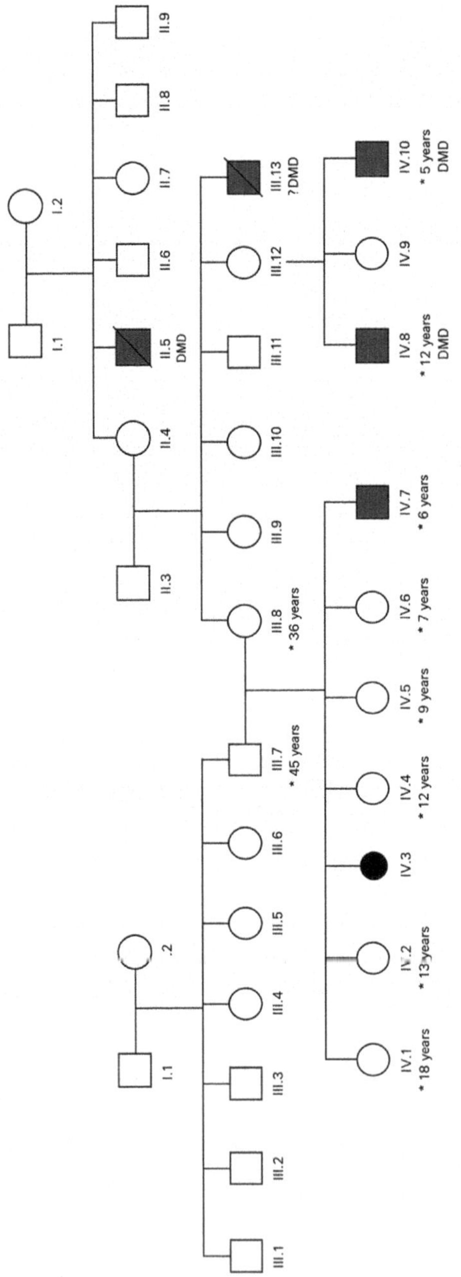

Three generation family pedigree for case 1.

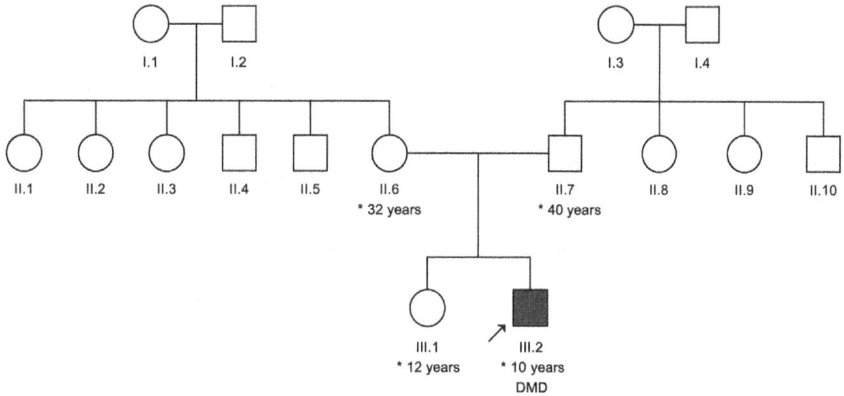

Three generation family pedigree for case 2.

17. Ethical and legal considerations in AS and DMD

The integration of genetic testing and precision diagnostics in rare disorders such as AS and DMD necessitates a careful appraisal of ethical and legal frameworks. While technological advancements have enabled earlier and more accurate diagnoses, they also raise significant concerns around consent, privacy, discrimination, and access particularly in a socio-culturally diverse nation like India.

17.1 Informed consent and genetic counseling

Informed consent forms the cornerstone of ethical genetic testing. For both AS and DMD, families must be counseled on the nature of testing, implications for health and reproduction, and the psychosocial impact of results.

- For pediatric cases, parental consent is often required, raising concerns about the child's future autonomy.
- Carrier and prenatal testing necessitate clear communication of inheritance patterns, recurrence risks, and reproductive options.

17.2 Privacy and confidentiality

Genetic data is sensitive and must be handled with strict confidentiality.

- Disclosure should be limited to authorized professionals.
- With increasing data sharing for research, families must be informed of storage protocols and future use, especially in the absence of comprehensive Indian data protection laws.

17.3 Genetic discrimination

India lacks dedicated legislation equivalent to the U.S. GINA Act, making individuals vulnerable to genetic discrimination in.

- **Employment**: Fear of bias in hiring or promotion decisions based on genetic risk.
- **Insurance**: Potential denial or premium loading based on carrier or affected status.
- **Marriage and Society**: Social stigma often leads to isolation, particularly for women, and may perpetuate myths around heredity and blame.

17.4 Prenatal testing and reproductive autonomy

The ethical landscape of prenatal testing is complex.

- Families may face moral dilemmas around termination following a positive diagnosis.
- Assisted reproductive technologies (e.g., PGD) raise questions of accessibility and ethical selection.
- Decisions must be supported by non-directive counseling those respects parental values and cultural norms.

17.5 Access to therapies and equity

Emerging therapies, such as gene therapy for DMD or targeted interventions in AS, bring forth ethical concerns about.

- **Equity**: High treatment costs may restrict access to affluent urban populations.
- **Clinical Trials**: Participation must be voluntary, informed, and ethically monitored, particularly in children.

17.6 End-of-life care and long-term planning

For progressive conditions like DMD, ethical considerations extend to advanced care planning, including ventilation support, palliative care, and respect for family preferences in terminal stages.

17.7 Legal gaps and policy needs

- **Absence of Specific Legislation**: India currently lacks robust legal instruments addressing genetic privacy and non-discrimination.
- **Disability Rights**: Though AS and DMD fall under the RPwD Act, implementation of inclusive education and healthcare remains inconsistent.
- **Pediatric Consent in Research**: Ethical oversight must balance parental authority with the child's emerging autonomy in trial participation.

Variant Interpretation and Incidental Findings: The increasing use of genomic technologies in disorders such as AS and DMD has introduced several ethical challenges related to variant interpretation, incidental findings, and disclosure practices. As testing methodologies transition from targeted assays to broader approaches such as whole-exome and whole-genome sequencing, clinicians and genetic counselors are increasingly confronted with complex results that may not always provide clear clinical answers.One major challenge involves the interpretation of variants of uncertain significance (VUS). A VUS refers to a genetic alteration for which current evidence is insufficient to determine whether the variant is pathogenic or benign. The identification of a VUS may create uncertainty for families, particularly in prenatal settings where reproductive decisions may be influenced by incomplete information. Misinterpretation of such findings can lead to unnecessary anxiety, inappropriate medical decisions, or inaccurate recurrence risk estimation. Therefore, VUS findings should be interpreted cautiously and discussed within the context of phenotype, family history, segregation analysis, and available functional evidence.Periodic re-evaluation of VUS findings is also important, as ongoing research and expanding genomic databases may result in future reclassification. Families should be counseled regarding the dynamic nature of genomic interpretation and the possibility that uncertain findings may later be upgraded or downgraded based on emerging evidence.Another important ethical concern involves incidental or secondary findings. During genomic testing performed for AS or DMD, unrelated pathogenic variants may occasionally be identified in genes associated with other medically actionable conditions, such as hereditary cancer syndromes or cardiac disorders. The disclosure of such findings raises important questions regarding patient autonomy, consent, psychological burden, and the clinician's duty to inform.Prior to testing, families should be informed about the possibility of incidental findings and given the opportunity to decide whether they wish to receive such information. This is particularly relevant in pediatric testing, where the disclosure of adult-onset conditions may have implications for the child's future autonomy and privacy. Ethical frameworks generally recommend prioritizing findings with established medical actionability and clear clinical benefit.The psychosocial impact of uncertain or unexpected findings should not be underestimated. Genetic counseling plays a central role in facilitating informed decision-making, supporting family understanding, and minimizing distress associated with ambiguous or incidental genomic information. As genomic testing becomes increasingly integrated into routine clinical care, the development of standardized policies for variant reporting, data sharing, reanalysis, and result disclosure will remain essential.

Ethical and legal issues surrounding AS and DMD in India are multidimensional, involving individual rights, societal norms, and systemic gaps. A framework emphasizing informed consent, non-discrimination, privacy, and equitable access is essential. Integration of bioethics in clinical practice, coupled with policy reform and public education, is vital to uphold the rights and dignity of affected individuals and their families.

18. Conclusion

AS and DMD represent two paradigmatic neurodevelopmental disorders, each with a distinct genetic etiology yet sharing profound implications for the child, family, and healthcare system. While AS is a neurogenetic disorder primarily affecting cognition, behavior, and communication due to abnormalities in maternal *UBE3A* gene expression, DMD is an X-linked muscular dystrophy with progressive motor decline and associated cognitive deficits caused by mutations in the DMD gene. Both conditions are severe, lifelong, and currently incurable, underscoring the critical importance of early and accurate diagnosis. With the advancement of genomic technologies, particularly NIPS, a new era in prenatal care has begun one that emphasizes early risk assessment, targeted molecular analysis, and informed reproductive choices. For families with a known history or identified carrier status, NIPS offers a safer, non-invasive alternative to traditional diagnostic methods. It facilitates the early detection of actionable genetic variants, thereby allowing time-sensitive decisions, including preimplantation genetic testing, in vitro fertilization, or early prenatal interventions. However, the clinical utility of NIPS in rare single-gene disorders like AS and DMD depends on robust pre-test counseling, validated testing platforms, and appropriate follow-up confirmatory diagnostics. Ethical considerations, such as the psychological impact of early diagnosis and potential stigmatization, must also be addressed within a framework of non-directive genetic counseling and patient autonomy. Ultimately, the integration of NIPS into routine prenatal screening protocols for high-risk pregnancies can transform the landscape of rare neurodevelopmental disorder management. It empowers families with knowledge, helps prevent recurrence through personalized reproductive planning, and reinforces the shift toward precision medicine in prenatal genetics. As our understanding of the human genome evolves, so too must our commitment to equitable access, multidisciplinary care, and ongoing research to improve outcomes for children affected by rare genetic disorders like AS and DMD.

List of abbreviations

AS	Angelman Syndrome
ASD	Autism Spectrum Disorder
CMA	Chromosomal Microarray
CNS	Central Nervous System
CNV	Copy Number Variation
DEL	Deletion
DNA	Deoxyribonucleic Acid
E6-AP	E6-Associated Protein
EEG	Electroencephalogram
FISH	Fluorescence In Situ Hybridization
GABA	Gamma-Aminobutyric Acid

GABA AR	GABA Type A Receptor
IC	Imprinting Center
ID	Intellectual Disability
MLPA	Multiplex Ligation-dependent Probe Amplification
MRI	Magnetic Resonance Imaging
NGS	Next-Generation Sequencing
NIPS	Non-Invasive Prenatal Screening
PCR	Polymerase Chain Reaction
PWS	Prader–Willi Syndrome
RNA	Ribonucleic Acid
SNP	Single Nucleotide Polymorphism
UBE3A	Ubiquitin Protein Ligase E3A
UPD	Uniparental Disomy

AI disclosure

During the preparation of this work the author(s) used ChatGpt, Gemini, Perplexity in order to assist with content structuring and language refinement. All material was reviewed and validated by the author for accuracy and originality. After using this tool/service, the author(s) reviewed and edited the content as needed and take(s) full responsibility for the content of the publication.

References

Aartsma-Rus, A., & Krieg, A. M. (2017). FDA approves eteplirsen for Duchenne muscular dystrophy: The next chapter in the eteplirsen saga. *Nucleic Acid Therapeutics, 27*(1), 1–3.

Birnkrant, D. J., Bushby, K., Bann, C. M., et al. (2018). Diagnosis and management of Duchenne muscular dystrophy, part 1: Diagnosis, and neuromuscular, rehabilitation, endocrine, and gastrointestinal and nutritional management. *The Lancet Neurology, 17* (3), 251–267.

Birnkrant, D. J., Bushby, K., Bann, C. M., et al. (2018). Diagnosis and management of Duchenne muscular dystrophy, part 2: Respiratory, cardiac, bone health, and orthopaedic management. *The Lancet Neurology, 17*(4), 347–361.

Chamberlain, J. S., Gibbs, R. A., Ranier, J. E., Nguyen, P. N., & Caskey, C. T. (1988). Deletion screening of the DMD locus via multiplex DNA amplification. *Nucleic Acids Research, 16* (23), 11141–11156. https://doi.org/10.1093/nar/16.23.11141

Chen, M., Chen, C., Huang, X., Sun, J., Jiang, L., Li, Y., Zhu, Y., Tian, C., Li, Y., Lu, Z., Wang, Y., Zeng, F., Yang, Y., Song, X., Peng, Z., Yin, C., & Chen, D. (2020). Noninvasive prenatal diagnosis for Duchenne muscular dystrophy based on direct haplotype phasing. *Prenatal Diagnosis, 40*(8), 918–924. https://doi.org/10.1002/pd.5641

Chitty, L. S., Mason, S., Barrett, A. N., McKay, F., & Lench, N. (2015). Non-invasive prenatal diagnosis for single-gene disorders. *Clinical Genetics, 87*(1), 25–35. https://doi.org/10.1111/cge.12505

Mendell, J. R., Rodino-Klapac, L. R., Sahenk, Z., et al. (2013). Eteplirsen for the treatment of Duchenne muscular dystrophy. *Annals of Neurology, 74*(5), 637–647.

Min, Y. L., Bassel-Duby, R., & Olson, E. N. (2019). CRISPR correction of Duchenne muscular dystrophy. *Annual Review of Medicine, 70*, 239–255.

Schouten, J. P., McElgunn, C. J., Waaijer, R., Zwijnenburg, D., Diepvens, F., & Pals, G. (2002). Relative quantification of 40 nucleic acid sequences by multiplex ligation-dependent probe amplification. *Nucleic Acids Research, 30*(12), e57. https://doi.org/10.1093/nar/gnf056

Williams, C. A., Driscoll, D. J., & Dagli, A. I. (2010). Clinical and genetic aspects of Angelman syndrome. *Genetics in Medicine, 12*(7), 385–395. https://doi.org/10.1097/GIM.0b013e3181def138

Further reading

Alexander, W. M., Smith, M., Freeman, B. J., Sutherland, L. M., Kennedy, J. D., & Cundy, P. J. (2013). The effect of posterior spinal fusion on respiratory function in Duchenne muscular dystrophy. *European Spine Journal, 22*(2), 411–416.

Aman, N. (Jan–Mar 2025). The role, importance of genetic counsellors and the awareness of genetic counselling in Indian scenario. *Journal of Human Reproductive Sciences, 18*(1), 45–47. https://doi.org/10.4103/jhrs.jhrs_34_25

Birnkrant, D. J., Bushby, K., Bann, C. M., Apkon, S. D., Blackwell, A., Brumbaugh, D., & Ward, L. M. (2018). Diagnosis and management of DMD, part 2: Respiratory, cardiac, bone health, and orthopaedic management. *The Lancet Neurology, 17*(4), 347–361. https://doi.org/10.1016/S1474-4422(18)30025-5

Bradley, W. G., Jones, M. Z., Mussini, J. M., & Fawcett, P. R. (1978). Becker-type muscular dystrophy. *Muscle & Nerve, 1*(2), 111–132.

Brooke, M. H., Fenichel, G. M., Griggs, R. C., Mendell, J. R., Moxley, R., & Miller, J. P. (1981). Clinical trial in duchenne dystrophy: I. The design of the protocol. *Muscle & Nerve, 4*(3), 186–197. https://doi.org/10.1002/mus.880040302

Bushby, K., Finkel, R., Birnkrant, D. J., Case, L. E., Clemens, P. R., Cripe, L., & DMD Care Considerations Working Group. (2010). Diagnosis and management of DMD, part 1: Diagnosis, and pharmacological and psychosocial management. *The Lancet Neurology, 9*(1), 77–93. https://doi.org/10.1016/S1474-4422(09)70271-6

Bushby, K. M., & Gardner-Medwin, D. (1993). The clinical, genetic and dystrophin characteristics of becker muscular dystrophy. I. Natural history. *Journal of Neurology, 240*(2), 98–104.

Carriero, P. L., Zangari, R., Sfreddo, E., Ghirardi, A., Schieppati, A., Barbui, T., & Biroli, F. (2024). Exploring the clinical and genetic landscape of AS: Patient-reported insights from an Italian registry. *Journal of Clinical Medicine, 13*(12), 3520. https://doi.org/10.3390/jcm13123520

Carvill, G. L., & Mefford, H. C. (2015). Advances in genetic testing and understanding of epilepsy. *Current Opinion in Pediatrics, 27*(6), 659–664. https://doi.org/10.1097/MOP.0000000000000272

Cernisova, Z., et al. (2023). Microdystrophin gene addition significantly improves muscle functionality and diaphragm muscle histopathology in a fibrotic mouse model of DMD. *International Journal of Biological Macromolecules, 264*, Article 130544. https://doi.org/10.1016/j.ijbiomac.2024.130544

Cheng, W., Wang, W., Li, C., Liu, Y., Han, J., Li, J., & Gao, L. (2020). Clinical application of non-invasive prenatal testing for DMD using cell-free fetal DNA in maternal plasma. *Prenatal Diagnosis, 40*(3), 351–360. https://doi.org/10.1002/pd.5608

de Visser, M., Bakker, E., Defesche, J. C., Bolhuis, P. A., & van Ommen, G. J. (1990). An unusual variant of becker muscular dystrophy. *Annals of Neurology, 27*(5), 578–581.

Doorenweerd, N., Straathof, C. S., Dumas, E. M., Spitali, P., Ginjaar, I. B., Schrans, D. G., & Kan, H. E. (2017). Relationships between lower extremity muscle fatty infiltration and cognitive functioning in DMD. *PLoS One, 12*(1), Article e0169323. https://doi.org/10.1371/journal.pone.0169323

Doriguzzi, C., Palmucci, L., Mongini, T., Chiado-Piat, L., Restagno, G., & Ferrone, M. (1993). Exercise intolerance and recurrent myoglobinuria as the only expression of Xp21 becker type muscular dystrophy. *Journal of Neurology, 240*(5), 269–271.

Duchenne, G. B. A. (1868). *Recherches sur la paralysie musculaire pseudo-hypertrophique ou paralysie myosclerosique*. Paris: J.-B. Baillière et fils.

Emery, A. E. H. (1982). Muscular dystrophy, benign X-linked (becker). In P. J. Vinken, & G. W. Bruyn (Eds.), *Handbook of clinical neurology* (pp. 86–88). Amsterdam: North Holland.

Emery, A. E. H. (2002). The muscular dystrophies. *The Lancet, 359*(9307), 687–695. https://doi.org/10.1016/S0140-6736(02)07815-7

Emery, A. E., & Skinner, R. (1976). Clinical studies in benign (becker type) xlinked muscular dystrophy. *Clinical Genetics, 10*(4), 189–201.

Eslahi, A., Alizadeh, F., Avan, A., et al. (2023). New advancements in CRISPR-based gene therapy of DMD. *Gene, 867*, Article 147358. https://doi.org/10.1016/j.gene.2023.147358

Finder, J. D., Birnkrant, D., Carl, J., Farber, H. J., Gozal, D., Iannaccone, S. T., & ATS Consensus Statement. (2004). Respiratory care of the patient with DMD: ATS consensus statement. *American Journal of Respiratory and Critical Care Medicine, 170*(4), 456–465. https://doi.org/10.1164/rccm.200307-885ST

Flanigan, K. M. (2014). Duchenne and becker muscular dystrophies. *Neurologic Clinics, 32*(3), 671–688. https://doi.org/10.1016/j.ncl.2014.05.002

Gupta, A., & Aman, N. (2025a). Embracing neurodiversity: The critical role of awareness and acceptance in autism SpectrumDisorders. *Arch Mol Biol Genet, 4*(1), 1–8.

Gupta, A., & Aman, N. (2025b). The genetic architecture of neurological disorders: An integrated review. *Egyptian Journal of Medical Human Genetics, 136*. https://doi.org/10.1186/s43042-025-00762-z

Hoffman, E. P., Brown, R. H., & Kunkel, L. M. (1987). Dystrophin: The protein product of the DMD locus. *Cell, 51*(6), 919–928. https://doi.org/10.1016/0092-8674(87)90579-4

Huguenard, J. R. (2023). Adult gene therapy for epilepsy in a model of AS: Hope or hype? *Epilepsy Currents, 23*(5), 312–314. https://doi.org/10.1177/15357597231191885

Jiang, Y. H., Pan, Y., Zhu, L., et al. (2010). Altered ultrasonic vocalization and impaired learning and memory in AS mouse model with a large maternal deletion from *UBE3A* to Gabrb3. *PLoS One, 5*(8), Article e12278. https://doi.org/10.1371/journal.pone.0012278

Kalia, S. S., Adelman, K., Bale, S. J., et al. (2017). Recommendations for reporting of secondary findings in clinical exome and genome sequencing. *Genetics in Medicine, 19*(2), 249–255.

Karpati, G., & Watters, G. V. (1980). Adverse anaesthetic reactions in duchenne dystrophy. *Excerpta Medica ICS, 527*, 206.

Kishino, T., Lalande, M., & Wagstaff, J. (1997). *UBE3A*/E6-AP mutations cause AS. *Nature Genetics, 15*(1), 70–73. https://doi.org/10.1038/ng0197-70

Lee, D., Chen, W., Kaku, H. N., et al. (2023). Antisense oligonucleotide therapy rescues disturbed brain rhythms and sleep in juvenile and adult mouse models of AS. *eLife, 12*, Article e81892. https://doi.org/10.7554/eLife.81892

Matsumura, T., et al. (1990). Cardiac involvement in DMD. *Brain & Development, 12*(4), 396–400. https://doi.org/10.1016/S0387-7604(12)80161-4

Matsumura, K., Tomé, F. M., Collin, H., Rieger, F., Leturcq, F., Jeanpierre, M., & Fardeau, M. (1990). Expression of dystrophin-associated proteins in dystrophin-positive muscle fibers (revertants) in DMD. *Neuromuscular Disorders, 1*(2), 125–132. https://doi.org/10.1016/0960-8966(90)90001-R

McDonald, C. M., Finkel, R. S., Flanigan, K. M., et al. (2024). Evaluating genetic modifiers of DMD disease progression using modeling and MRI. *Neurology, 102*(3), e123–e134. https://doi.org/10.1212/WNL.0000000000201163

McNally, E. M., & Kaltman, J. R. (2017). Heart failure in DMD: An emerging challenge. *Journal of the American College of Cardiology: Heart Failure, 5*(6), 512–514. https://doi.org/10.1016/j.jchf.2017.04.006

Mendell, J. R., Shilling, C., Leslie, N. D., Flanigan, K. M., Al-Dahhak, R., Gastier-Foster, J., & Weiss, R. B. (2012). Evidence-based path to newborn screening for DMD. *Annals of Neurology, 71*(3), 304–313. https://doi.org/10.1002/ana.23528

Meng, L., Ward, A. J., Chun, S., et al. (2015). Towards a therapy for AS by targeting a long non-coding RNA. *Nature, 518*(7539), 409–412. https://doi.org/10.1038/nature13975

Mercuri, E., & Muntoni, F. (2013). Muscular dystrophies. *The Lancet, 381*(9869), 845–860. https://doi.org/10.1016/S0140-6736(12)61897-2

Nathan, A., Ganesh, A., Godinez, R. I., Nicolson, S. C., & Greeley, W. J. (2005). Hyperkalemic cardiac arrest after cardiopulmonary bypass in a child with unsuspected Duchenne muscular dystrophy. *Anesthesia & Analgesia, 100*(3), 672–674.

Ozarkar, S. S., Patel, R. K. R., Vulli, T., et al. (2024). Comparative profiling of white matter development in the human and mouse brain reveals volumetric deficits and delayed myelination in AS. *Molecular Autism, 15*, 54. https://doi.org/10.1186/s13229-024-00636-y

Passamano, L., Taglia, A., Palladino, A., Viggiano, E., D'Ambrosio, P., Scutifero, M., & Nigro, G. (2012). Improvement of survival in DMD: Retrospective analysis of 835 patients. *Acta Myologica, 31*(2), 121–125.

Richards, S., Aziz, N., Bale, S., et al. (2015). Standards and guidelines for the interpretation of sequence variants: A joint consensus recommendation. *Genetics in Medicine, 17*(5), 405–424.

Ricotti, V., et al. (2016). Neurodevelopmental, emotional, and behavioural problems in DMD. *Developmental Medicine and Child Neurology, 58*(1), 77–84. https://doi.org/10.1111/dmcn.12922

Ringel, S. P., Carroll, J. E., & Schold, S. C. (1977). The spectrum of mild X-linked recessive muscular dystrophy. *Archives of Neurology, 34*(7), 408–416.

Rosenberg, H., & Heiman-Patterson, T. (1983). Duchenne's muscular dystrophy and malignant hyperthermia: Another warning [letter]. *Anesthesiology, 59*(4), 362.

Samaha, F. J., & Quinlan, J. G. (1996). Myalgia and cramps: Dystrophinopathy with wide-ranging laboratory findings. *Journal of Child Neurology, 11*(1), 214.

Sethna, N. F., Rockoff, M. A., Worthen, H. M., & Rosnow, J. M. (1988). Anesthesiarelated complications in children with Duchenne muscular dystrophy. *Anesthesiology, 68*(3), 462–465.

Specht, L., & Shapiro, F. (1990). Boys with duchenne and becker muscular dystrophy are clinically indistinguishable at presentation. *Annals of Neurology, 28*, 443 [abstr].

Tsagkaris, C., Papakosta, V., Miranda, A. V., et al. (2020). Gene therapy for AS: Contemporary approaches and future endeavors. *Current Gene Therapy, 19*(6), 359–366. https://doi.org/10.2174/1566523220666200107151025

Velasco, M. V., Colin, A. A., Zurakowski, D., Darras, B. T., & Shapiro, F. (2007). Posterior spinal fusion for scoliosis in Duchenne muscular dystrophy diminishes the rate of respiratory decline. *Spine, 32*(4), 459–465.

Willner, J., Nakagawa, M., & Wood, D. (1984). Drug-induced fiber necrosis in duchenne dystrophy. *The Italian Journal of Neurological Sciences, 5*(Suppl. 3), 117.

CHAPTER

7 Chromosomal deletion syndromes: DiGeorge syndrome and cri du chat syndrome

Syed Asra Kirmani, Mahrukh Hameed Zargar, Jaffar Abass Wani, Tahir Mohiuddin Malla, Wasia Showkat and Yasir Bashir

Advanced Centre of Human Genetics, Sher-i-Kashmir Institute of Medical Sciences, Srinagar, Jammu and Kashmir, India

1. Introduction

Aneuploidy and smaller chromosomal structural abnormalities like copy number variants (CNVs) (microdeletions or duplications), inversions, translocations and the formation of isochromosomes or ring chromosomes (Hsu, 1998) are among the chromosomal conditions that affect one in 150 live births (Nussbaum et al., 2007). Turner syndrome, triple X syndrome, Klinefelter syndrome and trisomy of chromosomes 13, 18 and 21 are some of the more common aneuploidies reported in live births. We concentrate on these in this chapter. The prognosis for aneuploidies survived greatly like individuals with trisomy 21 typically live into their 60s, while those with trisomy 13 or 18 hardly ever survive past their first year. Most of the other autosomal trisomy's result in early pregnancy losses. First-trimester spontaneous miscarriages are often caused by trisomy's, such as those of chromosome 16, unless there is mosaicism. In fetal screening, many aneuploidy cases are detected after cell-free DNA and/or sonographic results. During pregnancy, amniocentesis or chorionic villus sampling are used for confirmatory diagnostic testing (Vlachadis et al., 2023). In the postnatal period, a postnatal physical examination or the results of laboratory tests may raise suspicions of the diagnosis where blood is usually used for confirmatory testing. Standard karyotyping methods can detect chromosomal duplications and deletions that are typically larger than 5–10 megabases, representing significant gains or losses of genetic material.

CNVs, a broad category of structural alterations inside a chromosome that result in additional or missing copies of certain DNA sequences, are frequently detected using chromosomal microarrays. It can be categorized as benign, disease-causing, or of uncertain significance, and their clinical impact can differ widely depending on the individual and genetic context. The most common clinical characteristic of CNV is intellectual impairment (Rosenberg & Pascual, 2014). The most well-known and extensively documented microdeletion condition is DiGeorge syndrome,

Non-Invasive Prenatal Screening of Rare Fetal Genetic Diseases. https://doi.org/10.1016/B978-0-443-27660-6.00005-5

which is caused by a microdeletion at 22q11.2 (McDonald-McGinn & Sullivan, 2011). Patients with a pathogenic deletion of 22q11.2 show varying degrees of clinical severity; some need frequent hospitalization and medical treatment, while others don't need any clinical interventions. This also depends on the size and length of CNV and whether it is of parental origin. The best way to clarify these details is through chromosomal microarray. Copy number variation (CNV) testing can yield three classifications, one of which is "benign." This indicates that the individual's genetic copy number does not show any alterations associated with disease when compared to typical patterns found in the general population: pathological meaning that there is change in copy number when compared to general population and that is thought to be contributing factor to a disease or disability or variant of uncertain significance meaning that a change has been found when compared to general population but that the relationship between this change and disease is unclear (Kearney et al., 2011). Patients should receive counseling about their personal views and beliefs regarding possible consequences of having a child with disabilities if they are interested in getting tested for CNVS. Additionally, the notion that a variant may be carried by one of the parents and linked to a variable phenotype should be discussed in counseling (Fonda Allen et al., 2016).

The following is a general classification of chromosomal syndromes; breakpoint disruption syndromes, where only one or a few genes may be mutated; deletion syndromes, where a segment of the chromosome is missing; and duplication syndromes, where an extra segment of chromosome material is present. Deletion syndromes are more common than duplication syndromes. When normal genetic sequences are added or removed, a dosage effect occurs that results in an aberrant phenotype. The functional effect of having only one dosage of the genes in the remaining intact chromosome, which corresponds to the deleted segment, is referred to as "haploinsufficiency. A triple dose of genes due to duplication could be referred to by the opposite phrase, "triplo-excess." (Allen et al., 2016) Large-scale chromosomal deletions or duplications usually have a detrimental effect on fetal development until the conclusion of pregnancy. Down syndrome is a notable example of a condition where a distinct set of physical and developmental traits arises due to the presence of an extra copy of chromosome 21, a phenomenon known as trisomy 21. Postnatal survival is more commonly compatible with lesser anomalies. Numerous "classic" deletion disorders manifest as severe intellectual impairment and physical abnormalities.

As more precise cytogenetic techniques are developed, smaller deletions (microdeletions), linked to milder phenotypes are being found. While chromosomal analysis conducted as a screening investigation is necessary to identify the majority, some may be clinically identifiable by those with expertise in dysmorphology (Schinzel, 1994). Approximately 60 genes can be found in a single band, although there is significant variation, with both gene-dense and gene-sparse chromosomal regions. Chromosome 21, for example, is gene-sparse, makes up 1% of HAL, has 10 distinct bands in its long arm, and has only 225 genes. In certain cases, some larger regions were also mentioned, aggregation of cases with epilepsy was recorded. Few chromosome syndromes are commonly known to neurologists to be specifically linked to epilepsy. When evaluating a child with intellectual disability and noticeable physical differences, pediatric neurologists often request cytogenetic testing even without a clear suspicion of a specific genetic syndrome. In contrast, such chromosomal analyses are seldom part of the diagnostic process in adult epilepsy clinics, especially when the individual appears physically and cognitively typical. However, chromosomal abnormalities are found in about 6% of individuals who present with both epilepsy and intellectual disability. This figure increases significantly to nearly 50% in cases where seizures are accompanied by multiple congenital anomalies (Schinzel, 1994; Gardner & Sutherland, 1996, p. 375). People in these categories are frequently identified through distinct epilepsy syndromes, such as Lennox-Gastaut syndrome and West syndrome, which serve as clinical markers of underlying neurological or genetic conditions. Chromosome deletion syndromes are characterized by a unique set of phenotypic traits, including prenatal anomalies, growth and developmental delays, and neurocognitive impairment. Below is a detailed synopsis of a few selected syndromes.

2. Wolf-Hirschhorn syndrome

Wolf Hirschhorn syndrome (WHS) is a rare genetic disorder that results from a deletion of genetic material near the end of the short arm of chromosome 4. It has a female to male ratio of 2:1 and a prevalence of 1:20,000–50,000 births (Paradowska-Stolarz, 2014; Bailey, 2014). Intrauterine growth retardation, short stature, low weight, hypotonia, intellectual disability, epilepsy, and a particular craniofacial dysmorphism (also known as the "Greek warrior helmet") are among the distinctive characteristics of WHS. Skeletal abnormalities, immunological disorders, genitourinary tract defects, eye abnormalities, hearing loss, and congenital heart defects (pulmonary stenosis and septal defects) are additional related clinical manifestations (Cassidy, 1997). High anterior hairline with prominent glabella, wide nasal bridge that extends to the forehead, highly arched eyebrows, widely spaced eyes, epicanthus, short philtrum, downturned corners of the mouth and micrognathia are some of the suggestive craniofacial features. Most patients have low-set or posteriorly angulated ears with pits or tags, microcephaly and poorly developed ears (lobe-less pinnae, underdeveloped or missing cartilage) (Battaglia et al., 2015). A partial deletion of chromosome 4's distal short arm is typically the genetic defect, but complex chromosomal rearrangements such as translocations or ring chromosomes can also produce the WHS phenotype. Unbalanced chromosomal rearrangements may arise either spontaneously (de novo) or be inherited from a parent carrying a balanced translocation. Among the most frequently observed translocations are t(4p; 8p), along with others such as t(4p; 7p), t(4p; 11p), t(4p; 20q), t (4p; 21q), and t(4p; 12p). Additionally, structural anomalies like inverted duplications associated with terminal deletions on the same 4p arm, as well as unbalanced pericentric inversions, have also been reported (Battaglia et al., 2015; Nevado et al., 2020).

3. Prader–Willi syndrome

Prader–Willi syndrome (PWS) is a complex genetic condition that affects multiple bodily systems and is primarily linked to dysfunction of the hypothalamus. Key clinical features include reduced growth, excessive weight gain, underdeveloped sexual characteristics, cognitive and behavioral challenges, delayed development, and low muscle tone in infancy. The most significant contributors to health complications and early death in PWS are obesity and behavioral disturbances. This syndrome arises due to the absence of active paternal genes in a specific imprinted region on chromosome 15 (15q11-q13). This genetic deficiency can result from a deletion of the paternal chromosome, inheritance of two maternal copies (uniparental disomy), or errors in the imprinting process. Advances in genetic testing have enhanced the ability to detect these underlying causes early, allowing for timely intervention and management. Recent diagnostic improvements have also revealed subtle differences in physical and behavioral traits among individuals of various ethnic backgrounds and between those with deletions versus those with uniparental disomy. Additionally, individuals with a microdeletion in chromosome 20p13 may experience delays in diagnosis due to overlapping or less typical symptoms (Cassidy, 1997).

4. Jacobsen syndrome

Jacobsen syndrome is a rare genetic disorder caused by the partial deletion of the long arm of chromosome 11, specifically involving the 11q terminal region. It is classified as a contiguous gene syndrome affecting multiple systems, including motor and cognitive development. Fewer than 250 cases have been documented, with a reported prevalence of approximately one in 100,000 live births. The condition is observed more frequently in females than males, with a ratio of 2:1. Common clinical manifestations include impaired growth both before and after birth, delayed psychomotor development, and distinctive craniofacial features. These may include skull abnormalities, wide-set eyes (hypertelorism), drooping eyelids (ptosis), eye defects such as coloboma, downward-slanting eye openings, epicanthal folds, a broad nasal bridge, a short nose, a V-shaped mouth, and small, low-set ears that are rotated backward. Affected individuals often present with malformations in various organ systems, including the skeletal structure, brain and spinal cord, reproductive organs, gastrointestinal tract, kidneys, and heart. Additional complications may involve the sensory organs, immune function, and hormonal regulation. The chromosomal deletion typically spans 7–20 megabases and extends to the telomere, with the proximal breakpoint usually located at or beyond sub-band 11q23.3. In most cases (around 85%), the deletion occurs spontaneously (de novo), while the remaining 15% result from inherited chromosomal rearrangements or unbalanced segregation from a parent with a balanced translocation. A small subset of cases involves a breakpoint at the FRA11B fragile site.

Diagnosis is supported by clinical features such as thrombocytopenia, intellectual disability, and characteristic facial traits, alongside cytogenetic testing. Differential diagnoses include conditions like Turner syndrome, Noonan syndrome and acquired thrombocytopenia due to sepsis. Prenatal detection is possible through cytogenetic analysis using procedures such as amniocentesis or chorionic villus sampling. Comprehensive care requires a multidisciplinary team, including pediatric cardiologists, neurologists, ophthalmologists, and general pediatricians. Routine evaluations should include blood work, endocrine and immune system assessments, hearing tests, and ongoing follow-up. Infants with severe heart defects may need surgical intervention, and those with feeding difficulties might require nutritional support via feeding tubes. Hematological complications, particularly involving platelet function, demand close monitoring. Approximately 20% of affected infants die within the first 2 years, primarily due to congenital heart disease, while bleeding

Chromosomal Deletion syndrome	Chromosome Location	Incidence	Major Genes affected	Clinical features
Cri-du-chat	5p-	1: 15000to1: 50,000	Semaf, CTNND2	Microcephaly, high-pitched cry, severe psychomotor and intellectual disability (Cerruti Mainardi, 2006)
Wolf - hirschhorn	4p-	1:50,000	NSD2, MSX1, LETM1	Delayed growth and development, seizures, intellectual disability, characteristic facial features (Rjiba et al., 2021; Wiel et al., 2022).
Angelman	15q11-13	1:12,000 to 1:20,000	UBE3A	Seizures, problems with movement and balance, intellectual disability, delayed development, severe speech impairment (Bird, 2014).

Continued

Chromosomal Deletion syndrome	Chromosome Location	Incidence	Major Genes affected	Clinical features
Prader–Willi	15q11-13	1:10000to 1:3000	snoRNA genes (SNORD116 Cluster	Distinctive facial features, behavioral problems, constant hunger, obesity (Baldini et al., 2022).
Kleefstra	9q34.3	1:25000to 1:35,000	EHMT1	Speech impairment, intellectual disability, developmental delay (Ren et al., 2024)
22q11	22q11.2	1:4000	TBX1, COMT	Heart abnormalities, recurrent infections, distinctive facial Features, submucosal cleft palate (McDonald-McGinn et al., 2020)
Smith–Magenis	17p11.2	1:25,000	RAI1	distinctive facial features, Sleep disturbances, behavioral problems, delayed speech and language skills (Rinaldi et al., 2022)
Williams	7q11.23	1.7500 to 1:18,000	ELN, GTF2I, GTF2IRD1, LMK1	Distinctive facial features, unique personality characteristics, mild to moderate intellectual disability, cardiovascular problems (Khelifi et al., 2024).

complications are less frequent. Long-term survival rates and life expectancy beyond infancy remain uncertain (Mattina et al., 2009).

5. Cry du chat syndrome

Cri du Chat Syndrome (CdCS), first identified by Lejeune and colleagues in 1963, is a rare genetic condition (Lejeune et al., 1963) named after its hallmark symptom, a distinctive, high-pitched cry in infants that resembles the sound of a cat. This cry typically diminishes as the child grows older.

The disorder is caused by a deletion of genetic material on the short arm of chromosome 5 (5p), which may be either partial or complete. The size of the deletion often ranges between 35% and 55%, and in the majority of cases over 80% occurs spontaneously (de novo), without being inherited from a parent (Mainardi et al., 2001). Recent advancements in molecular genetics have significantly enhanced the ability to detect and understand CdCS. These technologies allow for more precise diagnosis and have contributed to a deeper understanding of the syndrome's variability and underlying genetic mechanisms.

6. Epidemiology

Cri du Chat Syndrome (CdCS), while classified as a rare autosomal genetic disorder, is among the more frequently occurring chromosomal syndromes in humans (Cornish & Pigram, 1996). Early epidemiological estimates placed their prevalence at approximately one in 50,000 live births, though subsequent studies have suggested it may be closer to one in 37,000 (Niebuhr, 1978a, 1978b). The condition appears slightly more common in females, but there is no significant evidence of variation in its occurrence across different racial or geographic populations.

CdCS accounts for less than 1% of individuals diagnosed with intellectual disabilities. In one study, Niebuhr identified the syndrome in roughly one out of every 350 Ind. among a group of 6000 with cognitive impairments. No consistent associations have been found between the syndrome and factors such as prenatal complications, parental age, or birth order within families (Cornish & Pigram, 1996)

A survey conducted by Cornish and Pigram (1996) involving 27 mothers of children with CdCS revealed that 37% had experienced pregnancy-related complications, and 44% reported difficulties during labor (Niebuhr, 1978a, 1978b). There have also been occasional reports linking maternal health issues such as anorexia, severe morning sickness, toxemia, and vaginal bleeding as well as parental exposure to radiation, with the occurrence of CdCS, though these findings remain anecdotal and inconclusive (Cornish & Pigram, 1996).

7. Clinical features

7.1 Neonatal period

One of the most recognizable features in infants with Cri du Chat Syndrome during the neonatal stage is their distinctive cry. This cry is typically described as high-pitched, weak, and monotonous, resembling the sound of a kitten. Although this vocal trait often fades within months or a few years, its duration can vary among individuals. In rare cases, some infants may not exhibit the characteristic cry at all or may instead be present with inspiratory stridor. Importantly, because similar vocalizations can occur in other neurological or genetic conditions, this symptom alone is not considered a definitive diagnostic marker for the syndrome (Cornish & Pigram, 1996).

Early case studies of Cri du Chat Syndrome (CdCS) (Mainardi et al., 2001) suggested that the characteristic cry might be linked to structural abnormalities in the larynx. These included features such as an underdeveloped and slender larynx with a diamond or quadrangular shape, a small and flexible epiglottis with a curved form, and unusual air space patterns in the posterior region during vocalization. However, subsequent investigations found that not all affected individuals exhibited these anatomical traits, indicating that other physiological or functional factors may also contribute to the vocal presentation (Colover et al., 1972; Howard, 1972).

In addition to the distinctive cry, infants with CdCS typically present with low birth weight-often falling below the 20th percentile (Baldini et al., 2022). Despite having a gestational age close to full term, these newborns tend to be smaller than average, consistently measuring below the 50th percentile in size, with reported average birth weights around 614 g (Colover et al., 1972).

Infants with Cri du Chat Syndrome often experience significant feeding difficulties, including weak sucking reflexes, trouble swallowing (dysphagia), low muscle tone (hypotonia), and issues such as gastroesophageal and nasal reflux. These challenges contribute to notably slow growth and developmental progress during the first 2 years of life (Colover et al., 1972).

While early studies focused on anatomical abnormalities in the larynx and epiglottis as contributing factors to the syndrome's hallmark symptoms, current understanding also implicates both structural and functional disruptions within the nervous system. These neurological factors may play a key role in feeding impairments and other developmental delays observed in affected individuals (Colover et al., 1972; Howard, 1972).

Infants with Cri du Chat Syndrome often face several serious health challenges early in life. Common complications include episodes of cyanosis, inspiratory stridor, and varying degrees of asphyxia (Rinaldi et al., 2022). Recurrent respiratory infections are frequently observed, and in some cases, pulmonary aspiration may lead to bronchiectasis. In a study by Niebuhr, there were 12 documented instances of acute asphyxia and 11 cases of complete asphyxia occurring shortly after birth.

Although seizures and muscle stiffness (spasticity) have occasionally been noted, full-blown epileptic episodes are relatively rare in this population. Alarmingly, the majority of neonatal deaths—around 90%—occur within the first year, with approximately 75% happening in the initial months. The leading causes of mortality include respiratory distress syndrome, congenital heart defects, and both typical and aspiration-related pneumonia (Colover et al., 1972).

7.2 Craniofacial manifestations

Individuals with Cri du Chat Syndrome commonly exhibit significant cognitive and developmental delays, often falling below the 50th percentile for mental function and below the 20th percentile for developmental milestones. Despite these challenges, current research does not indicate a consistent link between these impairments and structural abnormalities in the brain.

One of the hallmark physical traits in early childhood is microcephaly, which tends to be most pronounced during the first 2 years of life. Distinctive craniofacial features are also frequently observed, including a vertically oriented facial growth pattern, marked asymmetry, and a rounded or "moon-shaped" face—reported in over 80% of cases. As children grow older, these facial characteristics typically evolve, with the face becoming more elongated and narrow during adolescence and adulthood.

Epicanthal folds (90.2% of patients), downturned mouth corners (81%), large nasal bridges (87.2%), short philtrum (87.8%), convex facial profiles with mandibular micro retrognathia (96.7%), abnormal dermatoglyphics (92%) and premature gray hair are among the other facial changes commonly associated with CdCS (Howard, 1972).

7.3 Sensitive alterations

Multiple studies have identified sensory alterations in individuals with Cri du Chat Syndrome, particularly affecting hearing and vision. Among the most reported issues are strabismus (misalignment of the eyes), heightened sensitivity to sound, and discomfort in response to auditory stimuli. Less frequently observed complications include nearsightedness (myopia), cataracts, and unusual pupil responses such as increased sensitivity to methacholine and reduced responsiveness to pupil-dilating agents, which may be linked to underactivity of the dilator muscle. Auditory impairments often involve reduced hearing sensitivity, especially at higher frequencies, contributing to communication challenges and developmental delays in affected individuals (Takebayashi et al., 2006).

7.4 Other anomalies

Although not frequently observed, individuals with Cri du Chat Syndrome may present with abnormalities affecting the cardiovascular system, skeletal structure, genitourinary tract, metabolism, and immune function. Among the cardiac anomalies, the most reported are congenital defects such as patent ductus arteriosus combined with ventricular septal defects, and less often, associations with atrial septal defects. Additional features may include cyanosis, skin conditions like hemangiomas and cutis marmorata, and complex heart malformations such as tetralogy of Fallot. Musculoskeletal complications, particularly progressive scoliosis, are also frequently documented, highlighting the need for ongoing orthopedic evaluation in affected individuals (Niebuhr, 1978a, 1978b; Breg et al., 1970).

Muscular hypertonia may be a major factor in the development of this illness since patients with higher muscle tone demonstrated a noticeable worsening of scoliosis with growth (Breg et al., 1970). There are age-related variations in the genitourinary system. Youth cryptorchidism and adult hypogonadism are common (Martínez et al., 1993). Girls in their pubertal years have irregular menstrual cycles, menarche, and normal sexual secondary development (Niebuhr, 1978a, 1978b).

To date, only one documented case exists of two individuals with Cri du Chat Syndrome (CdCS) achieving pregnancy (Cornish & Bramble, 2002). Renal anomalies, while rare, have been observed and may include conditions such as ectopic kidneys, renal agenesis, or horseshoe kidney formation. In terms of metabolic function, some individuals may present with non-ketotic hyperglycemia or disruptions in purine synthesis, both of which could potentially impact neurological development. Although individuals with CdCS do not appear to have heightened susceptibility to infections, studies have noted alterations in immune cell behavior—specifically, changes in granulocyte activity and phagocytic function. These findings suggest subtle immune system involvement that may warrant further investigation.

7.5 Diagnosis and complementary tests

Since Cri du Chat Syndrome lacks a single definitive symptom, many of its observable traits such as a rounded facial appearance, widely spaced eyes, low-set ears, and downward-turned mouth corners—are often assessed subjectively. This variability can lead to diagnostic uncertainty and make it difficult to define a consistent clinical profile (Cerruti Mainardi, 2006). Therefore, recognizing a constellation of key features—such as microcephaly, distinct facial morphology, the characteristic high-pitched cry, and low birth weight—is more valuable for accurate diagnosis than relying on the search for a singular, pathognomonic sign.

Once Cri du Chat Syndrome (CdCS) is clinically suspected, the initial step in genetic evaluation typically involves karyotype analysis. If the karyotype appears normal but clinical features strongly suggest the syndrome, more refined molecular cytogenetic techniques such as fluorescence in situ hybridization (FISH), comparative genomic hybridization (CGH), or quantitative PCR are employed to detect smaller chromosomal deletions that standard karyotyping may miss.

FISH has significantly advanced the diagnostic process by enabling precise mapping of phenotypic traits to specific genomic regions. This technique has contributed to a surge in the identification of chromosomal deletion syndromes, including CdCS, and has reignited academic interest in understanding their genetic and clinical variability.

In recent years, fluorescence in situ hybridization (FISH) has been complemented by more advanced molecular techniques, notably comparative genomic hybridization (CGH). CGH utilizes DNA microarrays and quantitative PCR to analyze the genome at a higher resolution. One of its key advantages is the ability to scan the entire genome using overlapping probes, which enable the detection of subtle genetic alterations that traditional methods might miss. This comprehensive approach has significantly improved the identification of chromosomal imbalances and contributed to more accurate genetic diagnoses.

7.6 Prognosis and treatment

Although Cri du Chat Syndrome (CdCS) is associated with significant health challenges in infancy, many individuals live well into adulthood, with documented cases of patients reaching beyond 50 years of age. The highest risk of mortality occurs in the early stages of life—approximately 75% of reported deaths take place within the first few months after birth, and 90% occur during the first year (Cornish & Pigram, 1996; Niebuhr, 1978a, 1978b).

The prognosis is closely tied to the nature of chromosomal deletion, particularly its size and location on the short arm of chromosome 5. These genetic factors influence the severity of developmental delays, including cognitive and motor impairments (Niebuhr, 1978a, 1978b). Therefore, distinguishing between typical 5p deletions that result in the classic CdCS presentation and atypical deletions that produce a more variable clinical picture is essential for accurate diagnosis and management planning.

In cases of Cri du Chat Syndrome (CdCS), the clinical outcome is strongly influenced by the location and extent of the chromosomal deletion. When the deletion affects the critical region associated with the syndrome—typically spanning from band 5p15.2 to the terminal end of the short arm—patients tend to exhibit the full spectrum of symptoms, including significant cognitive and developmental impairments (Church et al., 1997). These are considered typical CdCS cases and are generally associated with a less favorable prognosis. Conversely, individuals with interstitial deletions that spare the key region often present with milder clinical features. These atypical cases are less likely to involve profound intellectual disability and may show improved developmental outcomes (McDonald-McGinn et al., 2015). Notably, when the deletion does not disrupt areas linked to the characteristic cry or language development, the overall prognosis tends to be more positive.

Patients with Cri du Chat Syndrome (CdCS) who carry imbalanced chromosomal translocations often present with more pronounced clinical symptoms and face a higher risk of mortality (Niebuhr, 1978a, 1978b). These genetic rearrangements tend to exacerbate the severity of the condition, particularly in the early stages of life.

One of the most impactful ways to improve outcomes for individuals with CdCS is through timely diagnosis. Early identification enables healthcare providers to implement targeted therapeutic strategies and preventive interventions that support both physical growth and cognitive development. By recognizing the syndrome early, clinicians can tailor care plans that optimize developmental potential and enhance overall quality of life.

7.7 DiGeorge syndrome

DiGeorge syndrome (DGS) refers to a collection of clinical features resulting from developmental disruptions in the pharyngeal arches during early embryonic growth. Although signs of the condition were noted as early as 1828, it was formally described by Dr.Angelo DiGeorge in 1965, highlighting a triad of immunodeficiency, hypoparathyroidism, and congenital heart defects. DGS is now recognized as part of a broader group of disorders associated with deletions on chromosome 22q11.2. This group includes conditions historically known by various names, such as Sedlackova syndrome, conotruncal anomaly face syndrome, and velocardio facial syndrome all of which share overlapping genetic and phenotypic characteristics (Fomin et al., 2010).

Although many syndromes share a common genetic origin, their diverse clinical presentations have historically led to the use of distinct names. This fragmented nomenclature has often complicated the diagnostic process for DGS, sometimes resulting in serious delays in identification and treatment (Cascella & Muzio, 2015). Recent studies advocate for the consolidation of these terms, recognizing them as interchangeable within the spectrum of 22q11.2 deletion syndromes.

DGS is characterized by underdevelopment or absence of the thymus, congenital heart anomalies, low calcium levels due to parathyroid hypoplasia, and immune system dysfunction. The most critical concern is the absence of thymic tissue, which plays a vital role in the maturation of T lymphocytes. Although complete thymic agenesis is rare, occurring in less than 1% of cases, it can lead to severe combined immunodeficiency. Immunological impairment in DGS varies, with some individuals experiencing mild to moderate deficiencies, while cardiac defects remain the most prevalent issue.

Additional features may include abnormalities in the oral cavity, kidneys, eyes, and gastrointestinal tract. Skeletal irregularities, psychiatric conditions, and developmental delays are also frequently observed. Cognitive outcomes vary among patients; rather than presenting with intellectual disability from birth, many exhibit a gradual decline in cognitive function over time (Cioffi et al., 2014).

7.8 Etiology

Approximately 90% of DGS cases are linked to a deletion on chromosome 22, specifically at the 22q11.2 region on the long arm. These deletions typically arise spontaneously, meaning they are not inherited and are absent in the genetic makeup of the affected individual's parents (Fomin et al., 2010). Within this chromosomal segment, researchers have identified around 90 genes, several of which have been studied using animal models.

Among these, the *T-box transcription factor 1* (TBX1) gene has received the most attention. Studies in mice have shown that disruptions in TBX1 lead to significant developmental defects in the heart, thymus and parathyroid glands, organs commonly affected in DGS. Moreover, TBX1 has been implicated in neuromicrovascular anomalies, which may contribute to the cognitive and behavioral challenges frequently observed in individuals with this syndrome (Cioffi et al., 2014; Paylor et al., 2006).

7.9 Pathophysiology

DGS arises from a microdeletion on chromosome 22 at the 22q11.2 locus, which encompasses over 90 genes (Zhang et al., 2006). Individuals affected by DGS display a broad spectrum of clinical features, with the most common being congenital heart defects, low calcium levels due to parathyroid hypoplasia, and an underdeveloped thymus. Among the genes located in this region, *TBX1* has been strongly linked to the core manifestations of the syndrome.

Disruption of *TBX1* interferes with the normal development of the pharyngeal pouches during embryogenesis, leading to structural deficiencies in the thymus and parathyroid glands. To explore the developmental mechanisms underlying these abnormalities, researchers have studied *TBX1* knockout models in both mice and zebrafish. In mice, the absence of *TBX1* results in pronounced defects in the pharyngeal region, heart, thymus, and parathyroid glands, along with signs of behavioral irregularities (Guner-Ataman et al., 2018). Similarly, zebrafish lacking *TBX1* exhibit malformations in the pharyngeal arches and thymus, as well as anomalies in ear development (Sumitomo et al., 2018).

Mouse models carrying the 22q11.2 deletion have been extensively studied, revealing molecular and behavioral traits that mirror those seen in neurological and psychiatric conditions such as Parkinson's disease, autism spectrum disorder, attention deficit hyperactivity disorder (ADHD), and schizophrenia (Meechan et al., 2015; McDonald-McGinn & Sullivan, 2011). These findings, when considered alongside the neurovascular abnormalities observed in *TBX1* knockout mice, suggest a biological foundation for the psychiatric symptoms often associated with DGS (Guner-Ataman et al., 2018; Sumitomo et al., 2018).

Notably, individuals with DGS face a significantly elevated risk estimated to be up to 30 times higher than the general population for developing schizophrenia. This strong association underscores the importance of early genetic screening and neurodevelopmental monitoring in patients with 22q11.2 deletions.

7.10 History and physical examination

Thorough history-taking and physical examination are essential components in identifying and evaluating DGS. Given the wide variability in clinical presentation, early suspicion based on developmental and physical findings can guide further diagnostic testing. While most cases are identified during prenatal screening or early childhood, some individuals may not receive a diagnosis until adulthood.

One of the earliest signs often noted by caregivers is delayed motor developments such as late achievement of milestones like rolling over or sitting independently. These motor delays frequently coincide with speech and language impairments, as well as learning difficulties. As children grow older, behavioral abnormalities may become more prominent, especially in individuals with a background of developmental challenges. In such cases, atypical behavior may be the primary clue leading to a diagnosis of DGS later in life (Fomin et al., 2010).

A comprehensive patient history plays a crucial role in identifying DGS. Key indicators may include.

- A known or suspected family history of DGS
- Relatives with abnormal results from genetic screening
- Developmental delays, particularly in motor milestones
- Behavioral concerns, cyanosis, or reduced exercise tolerance
- Frequent infections and speech difficulties linked to T-cell deficiency
- Poor growth or feeding challenges in infancy
- Neurological symptoms such as muscle spasms, twitching, tetany, or seizures

Physical examination can reveal a range of findings consistent with DGS.

- Cardiopulmonary assessment may uncover congenital heart defects, including aortic arch anomalies and conotruncal malformations such as tetralogy of Fallot, truncus arteriosus, pulmonary atresia with ventricular septal defect, transposition of the great arteries, interrupted aortic arch, or tricuspid atresia.
- Craniofacial evaluation may show features like cleft palate, widely spaced eyes (hypertelorism), ear malformations, and downward-slanting palpebral fissures
- Recurrent respiratory infections may be present due to impaired T-cell function from thymic underdevelopment
- Signs of hypocalcemia such as muscle twitching and spasm can result from parathyroid gland hypoplasia (Al-Sukaiti et al., 2010).

7.11 Treatment/management

- Effective management of DGS requires coordinated care across multiple medical specialties. Fortunately, many individuals with DGS exhibit only mild immune deficiencies, maintaining functional T-cell activity despite reduced production. Regular follow-up with an immunologist who specializes in primary immunodeficiencies is strongly advised to monitor immune status and guide treatment.
- In cases of complete DGS (cDGS), particularly in newborns, more aggressive interventions are necessary. These may include protective isolation, intravenous immunoglobulin (IVIG) therapy, prophylactic antibiotics, and in some cases,

transplantation of thymic tissue or hematopoietic stem cells to restore immune function.

- Immunization strategies including routine vaccines, boosters, IVIG, and antibiotic prophylaxis should be personalized based on each patient's lab results. Monitoring antibody levels to previously administered vaccines every 6–12 months helps determine whether re-vaccination is needed (Lackey & Muzio, 2023).
- There has been ongoing debate regarding the use of live vaccines such as measles-mumps-rubella (MMR), oral polio, and rotavirus in patients with DGS. However, recent studies support their safety and effectiveness in children over 1 year of age who demonstrate adequate immune function, specifically with CD8 counts above 300 and CD4 counts exceeding 500. It's important to note that the rotavirus vaccine has been associated with gastrointestinal complications in infants with severe combined immunodeficiency (SCID) and should be avoided in those with significantly reduced T-cell counts (Markert et al., 2008).
- **Cardiac Complications:** If congenital heart defects are not identified during prenatal imaging, they may manifest soon after birth as cyanotic heart disease, which can be life-threatening. In such cases, urgent evaluation by a pediatric cardiothoracic surgeon is essential. When transfusions are required, blood products should be irradiated, cytomegalovirus (CMV)-negative, and leukocyte-depleted to minimize the risk of transfusion-related graft-versus-host disease and reduce pulmonary complications, especially during procedures involving cardiopulmonary bypass.
- **Cleft Palate Management:** Children with cleft palate should be referred to specialists such as otolaryngologists, plastic surgeons, or oral and maxillofacial surgeons experienced in palatal reconstruction. Surgical correction not only enhances feeding and speech development but also decreases the likelihood of recurrent respiratory infections.
- **Autoimmune Surveillance:** Individuals with DiGeorge syndrome (DGS) are at increased risk for autoimmune conditions including immune thrombocytopenia (ITP), rheumatoid arthritis, autoimmune hemolytic anemia, Graves' disease, and Hashimoto's thyroiditis. Regular monitoring for autoimmune symptoms is crucial for early detection and management.
- **Hearing Evaluation:** Audiologic testing is vital for DGS patients, particularly young children who may not be able to express hearing difficulties. This is especially important in those with developmental delays or behavioral concerns, as undiagnosed hearing loss can exacerbate communication challenges.
- **Developmental Support:** Early intervention programs are highly beneficial for children with cognitive impairment. Speech therapy is especially important for those with craniofacial anomalies or intellectual disabilities that affect verbal communication.
- **Psychiatric Care:** Given the strong association between DGS and psychiatric conditions such as schizophrenia, individuals presenting with mood disturbances or psychotic symptoms should receive comprehensive mental health

evaluation and support (Meechan et al., 2015; McDonald-McGinn & Sullivan, 2011).

Genetic counseling is highly recommended for families affected by DGS, especially for parents considering future pregnancies or individuals with DGS who plan to have children. In cases where one parent carries the same chromosomal deletion found in their child, the likelihood of passing the condition to another offspring is approximately 50%, due to the autosomal dominant inheritance pattern of the 22q11.2 deletion.

7.12 Diagnosis

The diagnosis of 22q11.2 deletion syndrome, commonly known as DGS, relies heavily on immunological and clinical criteria. Immunophenotyping that reveals significant T-cell lymphopenia specifically, a $CD3^+$ T-cell count below 500 cells/mm^3 combined with at least two hallmark features can confirm the diagnosis. These features include conotruncal heart defects (such as truncus arteriosus, tetralogy of Fallot, interrupted aortic arch, or an aberrant right subclavian artery), persistent hypocalcemia requiring medical intervention, and genetic evidence of a 22q11.2 microdeletion.

Alternatively, a $CD3^+$ T-cell count under 1500 cells/mm^3 may also support the diagnosis when paired with either cytogenetic confirmation or one characteristic clinical sign such as congenital heart disease, calcium imbalance, or craniofacial anomalies. These criteria align with the European Society for Immunodeficiencies (ESID) classification system, which categorizes cases into definitive, probable, or possible DGS based on the combination of immunologic and phenotypic findings (Markert et al., 2009).

7.13 Prognosis

Complete DiGeorge syndrome (cDGS), the most severe manifestation of the 22q11.2 microdeletion, affects less than 1% of individuals with this genetic anomaly. The prognosis is extremely poor, particularly in infants who lack functional thymic tissue, leading to profound immunodeficiency. Without intervention such as thymic or hematopoietic stem cell transplantation, affected infants typically do not survive beyond their first year of life.

Even with transplantation, outcomes remain guarded. For example, in one study involving 50 infants who received thymic transplants, only 36 survived beyond 2 years, highlighting the complexity and risks associated with treatment (Markert et al., 2009). Factors contributing to mortality include pre-existing infections, complications from immunodeficiency, and other congenital anomalies. The prognosis for individuals with partial DGS varies widely and is largely influenced by the severity and combination of associated medical conditions. While some infants may succumb to complex congenital heart defects early in life, many others survive into adulthood with manageable symptoms. Due to its diverse presentation, DGS is believed to be significantly under-diagnosed particularly in adults who may exhibit only subtle congenital anomalies or mild cognitive and social challenges.

A considerable number of undiagnosed individuals likely live in the community with minimal support, as their symptoms may not prompt genetic evaluation. As genetic testing technologies continue to advance, it is anticipated that recognition and understanding of DGS will improve, enabling earlier diagnosis and more tailored care across the lifespan.

8. Conclusion

Chromosomal microdeletions, such as those seen in Cri du Chat syndrome and DiGeorge syndrome, illustrate how the absence of specific genetic segments can lead to complex disorders affecting multiple organ systems. Cri du Chat syndrome results from a terminal deletion on the short arm of chromosome 5, particularly at region 5p15.2. This genetic loss is associated with hallmark features including a distinctive high-pitched cry, cognitive impairment, facial anomalies, developmental delays, and occasionally congenital heart defects.

DiGeorge syndrome, on the other hand, stems from a microdeletion at chromosome 22q11.2, which spans roughly 30 to 50 genes. The clinical presentation is highly variable but commonly includes congenital heart malformations, cleft palate, underdeveloped thymus and parathyroid glands, T-cell deficiency, low calcium levels, developmental delays, and a heightened risk for psychiatric conditions such as schizophrenia.

Although the majority of cases in both syndromes estimated at 80%–90% are caused by spontaneous (de novo) deletions, inherited forms do occur. In DiGeorge syndrome, familial transmission accounts for up to 15% of cases, underscoring the importance of genetic counseling and family screening. Advancements in molecular diagnostics, including chromosomal microarray analysis, fluorescence in situ hybridization (FISH), multiplex ligation-dependent probe amplification (MLPA), and next-generation sequencing (NGS), have significantly enhanced the precision and speed of detection. These tools allow for earlier diagnosis and timely initiation of multidisciplinary interventions, which may include cardiac surgery, immunologic support, calcium supplementation, and tailored developmental therapies such as speech, occupational, and physical rehabilitation.

References

Al-Sukaiti, N., Reid, B., Lavi, S., Al-Zaharani, D., Atkinson, A., Roifman, C. M., & Grunebaum, E. (2010). Safety and efficacy of measles, mumps, and rubella vaccine in patients with DiGeorge syndrome. *J Allergy ClinImmunol, 126*(4), 868–869.

Allen, J. F., Stoll, K., & Bernhardt, B. A. (2016). Pre-and post-test genetic counseling for chromosomal and Mendelian disorders. In, *40. Seminars in perinatology* (pp. 44–55). WB Saunders.

Bailey, R. (2014). Wolf-hirschhorn syndrome: A case study and disease overview. *Advances in Neonatal Care, 14*(5), 318–321.

Baldini, L., Robert, A., Charpentier, B., & Labialle, S. (2022). Phylogenetic and molecular analyses identify SNORD116 targets involved in the Prader–Willi syndrome. *Molecular Biology and Evolution, 39*(1), Article msab348.

Battaglia, A., Carey, J. C., & South, S. T. (2015). Wolf–hirschhorn syndrome: A review and update. *American Journal of Medical Genetics Part C: Seminars in Medical Genetics, 169* (3), 216–223.

Bird, L. M. (2014). Angelman syndrome: Review of clinical and molecular aspects. *The application of clinical genetics*, 93–104.

Breg, W. R., Steele, M. W., Miller, O. J., Warburton, D., & DeCapoaA, A. P. W. (1970). The cri du chat syndrome in adolescents and adults:clinical finding in 13 older patients with partial deletion of the shortarm of chromosome No. 5(5p-). *J Pediatr, 77*, 782–791.

Cascella, M., & Muzio, M. R. (2015). Early onset intellectual disability in chromosome 22q11.2 deletion syndrome. *Revista Chilena de Pediatria, 86*(4), 283–286.

Cassidy, S. B. (1997). Prader-Willi syndrome. *Journal of Medical Genetics, 34*(11), 917–923.

Cerruti Mainardi, P. (2006). Cri du Chat syndrome. *Orphanet Journal of Rare Diseases, 1*(1), 33.

Church, D. M., Yang, J., Bocian, M., Shiang, R., & Wasmuth, J. J. (1997). A high resolution physical and transcript map of the Cri du chat region of human chromosome 5p. *Genome Research, 7*, 787–801.

Cioffi, S., Martucciello, S., Fulcoli, F. G., Bilio, M., Ferrentino, R., Nusco, E., & Illingworth, E. (2014). Tbx1 regulates brain vascularization. *Human Molecular Genetics, 23*(1), 78–89.

Colover, J., Lucas, M., Comley, J. A., & Roe, A. M. (1972). Neurologicalabnormalities in the 'cri-du-chat' syndrome. *Journal of Neurology Neurosurgery and Psychiatry, 35*, 711–719.

Cornish, K., & Bramble, D. (2002). Cri du chat syndrome: genotype-phenotype correlations and recommendations for clinical management. *Developmental Medicine and Child Neurology, 44*, 494–497.

Cornish, K. M., & Pigram, J. (1996). Developmental and behavioural characteristics of cri du chat syndrome. *Archives of Disease in Childhood, 75*(5), 448–450.

Fomin, A. B., Pastorino, A. C., Kim, C. A., Pereira, C. A., Carneiro-Sampaio, M., & Abe-Jacob, C. M. (2010). DiGeorge Syndrome: A not so rare disease. *Clinics, 65*(9), 865–869.

Fonda Allen, J., Stoll, K., & Bernhardt, B. A. (2016). Pre- and post-test genetic counseling for chromosomal and Mendelian disorders. *Seminars in Perinatology, 40*(1), 44–55. https://doi.org/10.1053/j.semperi.2015.11.007

Gardner, R. J. M., & Sutherland, G. R. (1996). *Chromosome abnormalities andgenetic counselling*. New York: Oxford University Press.

Guner-Ataman, B., González-Rosa, J. M., Shah, H. N., Butty, V. L., Jeffrey, S., Abrial, M., Boyer, L. A., Burns, C. G., & Burns, C. E. (2018). Failed Progenitor specification underlies the cardiopharyngeal phenotypes in a zebrafish model of 22q11.2 deletion syndrome. *Cell Reports, 24*(5), 1342–1354.e5.

Howard, R. O. (1972). Ocular abnormalities in the cri du chat syndrome. *American Journal of Ophthalmology, 73*, 949–954.

Hsu, L. Y. (1998). Prenatal diagnosis of chromosomal abnormalities through amniocentesis. *Genetic Disorders and the Fetus, 4*, 179–180.

Kearney, H. M., Thorland, E. C., Brown, K. K., Quintero-Rivera, F., & South, S. T. (2011). Working Group of the American College of Medical Genetics Laboratory Quality

Assurance Committee. American College of Medical Genetics standards and guidelines for interpretation and reporting of postnatal constitutional copy number variants. *Genetics in Medicine, 13*, 680–685.

Khelifi, R., Jelloul, A., Ajmi, H., Slimani, W., Dimassi, S., Rjiba, K., … Mougou-Zerelli, S. (2024). Toward autism spectrum disorders and Williams-Beuren syndrome co-occurrence condition in Tunisian patients: Genetic insights. *AIMS Molecular Science, 11*(4), 379–394.

Lackey, A. E., & Muzio, M. R. (2023). *DiGeorge syndrome*. StatPearls Publishing. InStat-Pearls [Internet].

Lejeune, J., Lafourcade, J., Berger, R., Vialatte, J., Boeswillwald, M, S. P., et al. (1963). [3 cases of partial deletion of the short arm of a 5 chromosome. *Comptes Rendus Hebdomadaires des Seances de l'Academie des Sciences, 257*, 3098–3102.

Mainardi, P. C., Perfumo, C., Calì, A., Coucourde, G., Pastore, G., Cavani, S., et al. (2001). Clinical and molecular characterisation of 80 patientswith 5p deletion: Genotype-phenotype correlation. *Journal of Medical Genetics, 38*, 151–158.

Markert, M. L., Devlin, B. H., Chinn, I. K., & McCarthy, E. A. (2008). Thymus transplantation in complete Digeorge Anomaly. *Immunologic Research, 44*, 61–70.

Markert, M. L., Devlin, B. H., Chinn, I. K., & McCarthy, E. A. (2009). Thymus transplantation in complete DiGeorge anomaly. *Immunologic Research, 44*(1–3), 61–70.

Martínez, J. E., Tuck-Muller, C. M., & Superneau, D. (1993). Wertelecki, W. Fertility and the cri du chat syndrome. *Clinical Genetics, 43*, 212–214.

Mattina, T., Perrotta, C. S., & Grossfeld, P. (2009). Jacobsen syndrome. *Orphanet Journal of Rare Diseases, 4*(1), 9.

McDonald-McGinn, D. M., Hain, H. S., Emanuel, B. S., & Zackai, E. H. (2020). 22q11. 2 Deletion Syndrome.

McDonald-McGinn, D. M., & Sullivan, K. E. (2011). Chromosome 22q11. 2 deletion syndromes (DiGeorge syndrome/velocardiofacial syndrome). *Medicine, 90*(1), 1–18.

McDonald-McGinn, D. M., & Sullivan, K. E. (2011). Chromosome 22q11.2 deletion syndrome (DiGeorge syndrome/velocardiofacial syndrome). *Medicine (Baltimore), 90*(1), 1–18.

McDonald-McGinn, D. M., Sullivan, K. E., Marino, B., Philip, N., Swillen, A., Vorstman, J. A., Zackai, E. H., Emanuel, B. S., Vermeesch, J. R., Morrow, B. E., Scambler, P. J., & Bassett, A. S. (2015). 22q11.2 deletion syndrome. *Nature Reviews Disease Primers, 1*, Article 15071.

Meechan, D. W., Maynard, T. M., Tucker, E. S., Fernandez, A., Karpinski, B. A., Rothblat, L. A., & LaMantia, A. S. (2015). Modeling a model: Mouse genetics, 22q11.2 Deletion Syndrome, and disorders of cortical circuit development. *ProgNeurobiol, 130*, 1–28.

Nevado, J., Ho, K. S., Zollino, M., Blanco, R., Cobaleda, C., Golzio, C., & Lapunzina, P. (2020). International meeting on wolf-hirschhorn syndrome: Update on the nosology and new insights on the pathogenic mechanisms for seizures and growth delay. *American Journal of Medical Genetics, Part A, 182*(1), 257–267.

Niebuhr, E. (1978a). The Cri du Chat syndrome: epidemiology, cytogeneticsand clinical features. *Human Genetics, 44*, 227–275.

Niebuhr, E. (1978b). Cytologic observations in 35 individuals with a 5pkaryotype. *Human Genetics, 42*, 143–156.

Nussbaum, R. L., McInnes, R. R., Willard, H. F., et al. (2007). *Thompson & Thompson genetics in medicine* (7th.). Philadelphia: Saunders/Elsevier.

Paradowska-Stolarz, A. M. (2014). Wolf-Hirschhorn syndrome (WHS)-literature review on the features of the syndrome. *Advances in Clinical and Experimental Medicine, 23*(3), 485–489.

Paylor, R., Glaser, B., Mupo, A., Ataliotis, P., Spencer, C., Sobotka, A., Sparks, C., Choi, C. H., Oghalai, J., Curran, S., Murphy, K. C., Monks, S., Williams, N., O'Donovan, M. C., Owen, M. J., Scambler, P. J., & Lindsay, E. (2006). Tbx1 haploinsufficiency is linked to behavioral disorders in mice and humans: Implications for 22q11 deletion syndrome. *Proc Natl Acad Sci U S A, 103*(20), 7729–7734.

Ren, R., Liu, Y., Liu, P., Zhao, J., Hou, M., Li, S., … Yuan, A. (2024). Clinical characteristics and genetic analysis of four pediatric patients with Kleefstra syndrome. *BMC Medical Genomics, 17*(1), 290.

Rinaldi, B., Villa, R., Sironi, A., Garavelli, L., Finelli, P., & Bedeschi, M. F. (2022). Smith-magenis syndrome—clinical review, biological background and related disorders. *Genes, 13*(2), 335.

Rjiba, K., Ayech, H., Kraiem, O., Slimani, W., Jelloul, A., Ben Hadj Hmida, I., … Mougou-Zerelli, S. (2021). Disorders of sex development in Wolf–Hirschhorn syndrome: A genotype–phenotype correlation and MSX1 as candidate gene. *Molecular Cytogenetics, 14*(1), 12.

Rosenberg, R. N., & Pascual, J. M. (2014). *Rosenberg's molecular and genetic basis of neurological and psychiatric disease* (5th ed.). Elsevier Science.

Schinzel, A. (1994). *Human cytogenetics database*. Oxford: Oxford University. Press (CD ROM version).

Sumitomo, A., Horike, K., Hirai, K., Butcher, N., Boot, E., Sakurai, T., Nucifora, F. C., Bassett, A. S., Sawa, A., & Tomoda, T. (2018). A mouse model of 22q11.2 deletions: Molecular and behavioral signatures of Parkinson's disease and schizophrenia. *Science Advances, 4*(8), Article eaar6637.

Takebayashi, T., Obata, H., Minaki, Y., Sekine, M., Imoto, K., Yokogushi, K., et al. (2006). Scoliosis in cat cry syndrome. *Journal of Orthopaedic Science, 11*, 259–263.

Vlachadis, N., Papadopoulou, T., Vrachnis, D., Manolakos, E., Loukas, N., Christopoulos, P., Pappa, K., & Vrachnis, N. (2023). Incidence and types of chromosomal abnormalities in first trimester spontaneous miscarriages: A Greek single-center prospective study. *Maedica, 18*(1), 35.

Wiel, L. C., Bruno, I., Barbi, E., & Sirchia, F. (2022). From Wolf-Hirschhorn syndrome to NSD2 haplo insufficiency: A shifting paradigm through the description of a new case and a review of the literature. *Italian Journal of Pediatrics, 48*(1), 72.

Zhang, Z., Huynh, T., & Baldini, A. (2006). Mesodermal expression of Tbx1 is necessary and sufficient for pharyngeal arch and cardiac outflow tract development. *Development, 133* (18), 3587–3595.

CHAPTER

Rare metabolic disorders: Alagille syndrome and hypophosphatasia

8

Md. Rashid Alam[1], Onaiza Ansari[1], Syeda Rushna[1], Mohd Shariq[2], Javaid Ahmad Sheikh[1]

[1]*Department of Biotechnology, Jamia Hamdard, New Delhi, India;* [2]*GITAM School of Science, Department of Life Sciences, GITAM University Hyderabad, Hyderabad, Telangana, India*

1. Introduction

Rare diseases are defined by low prevalence, but thresholds vary by jurisdiction; the European Union definition is <5 per 10,000 people, while national cutoffs span roughly 5–80 per 10,000 (Moliner & Waligora, 2017; Nguengang Wakap et al., 2020). Despite their individual rarity, Orphanet-based analyses estimate that rare diseases affect roughly 3.5%–5.9% of people worldwide (Nguengang Wakap et al., 2020). Phenotypic heterogeneity and limited clinical familiarity can delay molecular diagnosis and coordinated multidisciplinary care. Alagille syndrome (ALGS) is a prototypical autosomal dominant multisystem disorder caused by disrupted Notch signaling. Pathogenic variants are identified predominantly in JAG1 (~94% of clinically defined cases) and less often in NOTCH2 (~2–4%), with a small residual fraction remaining genetically unresolved (Gilbert et al., 2019; Kohut et al., 2021). Population prevalence remains uncertain because mild phenotypes are underascertained; published estimates span historical figures near ~1 in 70,000 live births and genotype-informed projections closer to ~1 in 30,000 (Vandriel et al., 2023). Clinically, ALGS combines cholestasis and bile duct paucity with variably penetrant cardiac, skeletal, ocular, renal, and vascular manifestations, and severity can vary markedly even among relatives (Ayoub & Kamath, 2020).

Hypophosphatasia (HPP) is a rare metabolic bone disorder that results from loss-of-function variants in the ALPL gene, which encodes tissue-nonspecific alkaline phosphatase (TNSALP). When TNSALP activity is reduced, inorganic pyrophosphate accumulates outside cells and other substrates (e.g., pyridoxal-5′-phosphate), leading to impaired hydroxyapatite formation and a phenotype ranging from perinatal lethal hypomineralization to childhood rickets/craniosynostosis, premature loss of deciduous teeth, and adult fractures with chronic musculoskeletal pain (Seefried et al., 2025; Tournis et al., 2021). Severe early-onset disease has been classically estimated at ~1/100,000 births, while frequencies of milder forms vary widely across molecular and laboratory-based studies; prevalence is therefore best given as ranges (Mornet, 2007; Mornet et al., 2011). A key diagnostic pitfall is

Non-Invasive Prenatal Screening of Rare Fetal Genetic Diseases. https://doi.org/10.1016/B978-0-443-27660-6.00002-X

failure to recognize persistently low serum alkaline phosphatase, particularly in adults labeled with "osteoporosis" (Seefried et al., 2025).

Prenatal genomics provides opportunities for earlier risk stratification. Cell-free fetal DNA (cffDNA) was first demonstrated in maternal plasma in 1997, enabling non-invasive prenatal screening (NIPS) by cfDNA sequencing, typically offered from ~10 weeks' gestation (Lo et al., 1997). At 11–13 weeks, fetal fraction has a median of ~10% (IQR 7.8%–13.0%) but varies substantially and can drive "no-call" results at low fraction (Ashoor et al., 2013). Professional guidance emphasizes that cfDNA is a screening test; screen-positive results should be confirmed with diagnostic testing (chorionic villus sampling or amniocentesis), which carries a small procedure-related pregnancy loss risk (estimated ~0.1–0.3%, depending on study design and baseline risk) (ACOG, 2020; Akolekar et al., 2015). Beyond aneuploidy, cfDNA analysis is being extended to selected monogenic disorders via targeted and dosage/haplotype-based methods; early clinical series report the highest yields in the presence of family history or fetal ultrasound anomalies, but performance metrics and counseling needs remain less certain for rare conditions (Hanson et al., 2022; Mohan et al., 2022; Vora et al., 2024). Together, ALGS and HPP illustrate why earlier, safer prenatal risk assessment for monogenic disease is clinically meaningful.

2. Rare metabolic disorders—Alagille syndrome and hypophosphatasia

Alagille syndrome, also known as arteriohepatic dysplasia, is a multisystem disorder that involves several organs at once, and its signs and symptoms vary greatly from person to person. It was first described in the early 1970s (Ayoub & Kamath, 2022). It follows an autosomal dominant pattern of inheritance like Huntington's disease (Krantz et al., 1997). Offspring of people with ALGS have a 50% chance of inheriting it. When we look at the distribution of inherited variants, 40% of people have inherited them, while the remaining 60% have a newly mutated pathogenic variant. This is because of heterozygous mutations in either the *JAG1* gene with 26 exons or the *NOTCH2* gene with 34 exons. Pathogenic JAG1 variants account for most clinically defined cases, up to about 94% in intensively screened cohorts, with a much smaller proportion attributable to NOTCH2 variants. Notch signaling is a highly conserved cell–cell communication pathway that serves a central role in embryonic patterning and organogenesis, including the development of the biliary tree, heart, vasculature, skeleton, and eye. In this pathway, the extracellular domain of the Jagged1 ligand on 1 cell binds to the Notch2 receptors on an adjacent cell, triggering a series of proteolytic cleavages that release the Notch intracellular domain. The cascade then translocates to the nucleus to modulate transcription of downstream target genes and thereby influence cell fate decisions (Ahn et al., 2015; Gilbert et al., 2019). Comprehensive mutation surveys have cataloged hundreds of distinct

JAG1 variants in ALGS, encompassing frameshift and nonsense variants, splice-site changes, and exon-level or whole-gene deletions, as depicted in Fig. 8.1, most of which are predicted to cause haploinsufficiency. A few whole gene deletions have also been found. Owing to the high variable expressivity of the genes, clinical manifestations can't be predicted by prenatal genetic testing (Table 8.1).

HPP is a rare inherited metabolic bone disorder caused by deficient activity of TNSALP, encoded by the ALPL gene. TNSALP is the predominant alkaline phosphatase isoenzyme in bone, liver, and kidney, and accounts for the great majority of total serum alkaline phosphatase activity. TNAP accounts for 95% of total ALP activity (Mornet, 2008). The disease manifests itself with defective bone and tooth mineralization. Symptoms are varied, ranging from still births without mineralized bones to early deciduous tooth loosening without any bone symptoms (Khan et al., 2024). Transmission is autosomal recessive for the severe kind but can be either autosomal dominant or recessive for the moderate kind (Mornet, 2017). Since the clinical manifestations are similar to many other disorders, the median time for diagnosis is 5.7 years (Seefried et al., 2020). The symptoms can be managed through enzyme replacement therapy. Table 8.2 provides a detailed list of research groups involved in studying the biology of HPP, along with their key contribution.

3. Clinical features and manifestations

3.1 Alagille syndrome

ALGS is a multisystem disorder with striking phenotypic variability, even within the same family. In the absence of a confirmed molecular diagnosis, clinical diagnostic criteria are typically met when characteristic findings are present in at least three of seven organ systems (liver, heart, skeleton, eye, face, kidney, and vasculature) (as represented in Fig. 8.2), after exclusion of alternative causes of cholestasis (Saleh et al., 2016). Hepatic involvement usually dominates the early clinical picture. Around 80%–90% of affected infants present with neonatal cholestasis and conjugated hyperbilirubinemia, typically accompanied by elevated gamma-glutamyl transferase (GGT) activity. Chronic cholestasis leads to fat-soluble vitamin deficiency, poor growth, fractures, and intense pruritus, which together affect a large majority of children. Hypercholesterolemia with cutaneous or tendinous xanthomas occurs in a substantial subset, reflecting impaired biliary cholesterol excretion. Enlargement of the spleen (splenomegaly) may manifest in approximately 66% of patients. Chronic cholestasis (inability of bile to flow normally from the liver to the duodenum) may result in FSVD (fat-soluble vitamin deficiency), poor growth, fractures, and pruritus (seen in roughly 74% of patients). It can also lead to hypercholesterolemia and the development of xanthomas (localized lipid deposits within organ systems) in approximately 24% of patients. This can lead to end-stage liver disease in upto 15% of cases. ALGS affects the cardiac system in 94% of the patients, marking the highest penetrance (Fig. 8.3). It may manifest as congenital cardiac defects, such as Tetralogy of Fallot

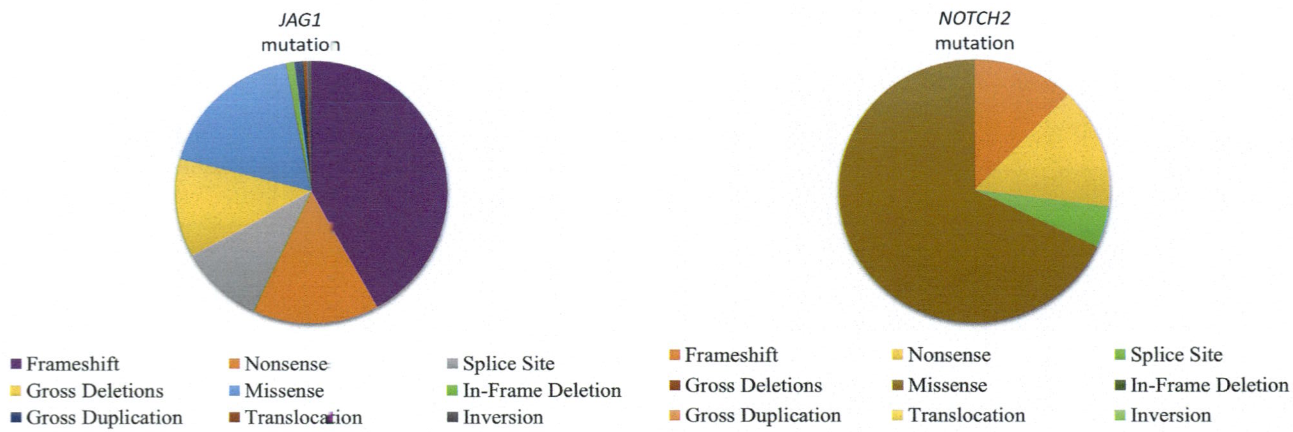

FIGURE 8.1

Alagille syndrome mutation update: Comprehensive overview of JAG1 and NOTCH2 mutation frequencies and insight into missense variant classification.

Table 8.1 Key research contributions to the clinical characterization of Alagille syndrome by organ system, phenotypic feature, and genotype-phenotype relationship.

Primary author/group	Key contributions	Years
Alagille et al. (1975)	Original syndrome description; bile duct paucity; characteristic facies; cardiac involvement	1969–1975
Emerick et al. (1999) Emerick et al. (2005)	Pulmonic stenosis prevalence; TOF; neurovascular accidents; growth failure; developmental delays; splenomegaly; mortality causes	1999, 2005
Kamath et al. (2002) Kamath et al. (2003) Kamath et al. (2004) Kamath et al. (2015)	Family variability; craniosynostosis; vascular anomalies; growth failure; mortality; intracranial vascular malformations	2002, 2003, 2004, 2015
McElhinney et al. (2002)	Cardiac manifestations (90%–97%); posterior embryotoxon; vertebral anomalies	2002
Kohut et al. (2021)	Pulmonary vascular abnormalities (94%); renal manifestations (39%); comprehensive phenotype review	2021
Vandriel et al. (2023)	Neonatal cholestasis (85%); pruritus timing; xanthomas; liver transplant outcomes; mortality rate (8.5%)	2023
Hingorani et al. (1999)	Axenfeld/Rieger anomaly; retinal pigmentary changes (32%)	1999
El-Koofy et al. (2011)	Optic disk drusen (90% on ultrasound)	2011
Da Palma et al. (2021)	Liver disease prevalence in non-hepatology cohort; peripheral chorioretinal changes	2021
Gilbert et al. (2019)	>700 individuals with pathogenic variants; comprehensive mutation update	2019
Loomes et al. (2019)	Bone mineral deficiency quantification	2019
Mouzaki et al. (2016)	Prognostic bilirubin threshold	2016
Schindler et al. (2021)	Hepatocellular carcinoma risk	2021
Salem et al. (2012)	Adult renal manifestations (hypertension, renal artery stenosis)	2012
Leung et al. (2022)	Attention and executive function impairment	2022
Paez-Escamilla et al. (2022)	Progressive vision loss; macular atrophy	2022
Reynal et al. (2023)	Oral and dental manifestations	2023
Lin et al. (2012)	Cross-cultural facial feature assessment	2012
Woolfenden et al. (1999), Rocha et al. (2012)	Moyamoya disease association	1999, 2012
Tretter and McElhinney (2018)	Cardiac defect frequency hierarchy	2018

Continued

Table 8.1 Key research contributions to the clinical characterization of Alagille syndrome by organ system, phenotypic feature, and genotype-phenotype relationship.—*cont'd*

Primary author/group	Key contributions	Years
Turnpenny and Ellard (2012)	Delayed puberty; high-pitched voice	2012
Mitchell et al. (2018)	Clinical variability spectrum	2018
Rodrigues Bento et al. (2022)	Isolated aneurysmal disease	2022

ALGS, *Alagille syndrome;* TOF, *tetralogy of fallot.*

Table 8.2 Key research contributions to the clinical characterization of Hypophosphatasia by subtype, treatment outcome, and genotype-phenotype relationship.

Primary author/group	Key contributions	Years
Huggins et al. (2020)	Intrafamilial variability; sibling concordance	2020
Duffus et al. (2018)	Unfavorable ERT outcomes possible	2018
Whyte et al. (2016) Whyte et al. (2019)	Asfotase alfa improves survival (perinatal/infantile)	2016, 2019
Wenkert et al. (2011)	Perinatal benign—skeletal manifestations resolve	2011
Schroth et al. (2021)	Early ERT improves oral health	2021
Whyte et al. (2016)	Gait/walk test/step length improve with ERT (juvenile)	2016
Pierpont et al. (2021)	ADHD and sleep disturbances (mild childhood HPP)	2021
Seefried et al. (2020)	Adult HPP: bone pain, non-skeletal disease, quality of life	2020
Dahir et al. (2022)	Adult HPP: muscle weakness, dental problems, quality of life	2022
Mori et al. (2016)	Odontohypophosphatasia progression to other HPP forms	2016
Mornet et al. (2021)	Heterozygous variants: osteoporosis, fracture risk, PLP/PEA predictors	2021
Fauvert et al. (2009)	Residual enzymatic activity variants; dominant-negative effect	2009
Lira Dos Santos et al. (2025)	Dentoalveolar complex histopathology	2025

ADHD, *attention-deficit/hyperactivity disorder;* ERT, *enzyme replacement therapy;* HPP, *hypophosphatasia;* PEA, *phosphoethanolamine;* PLP, *pyridoxal-5′-phosphate.*

(16%). It may also manifest as stenosis of the valvular/supravalvular aorta or of the peripheral pulmonary artery (67%). A distinct inverted triangular-shaped face with a pointed chin and a broad nasal bridge is seen in patients. They also showcase a broad forehead, prominent ears, and deep-set eyes with hypertelorism. Ophthalmologic

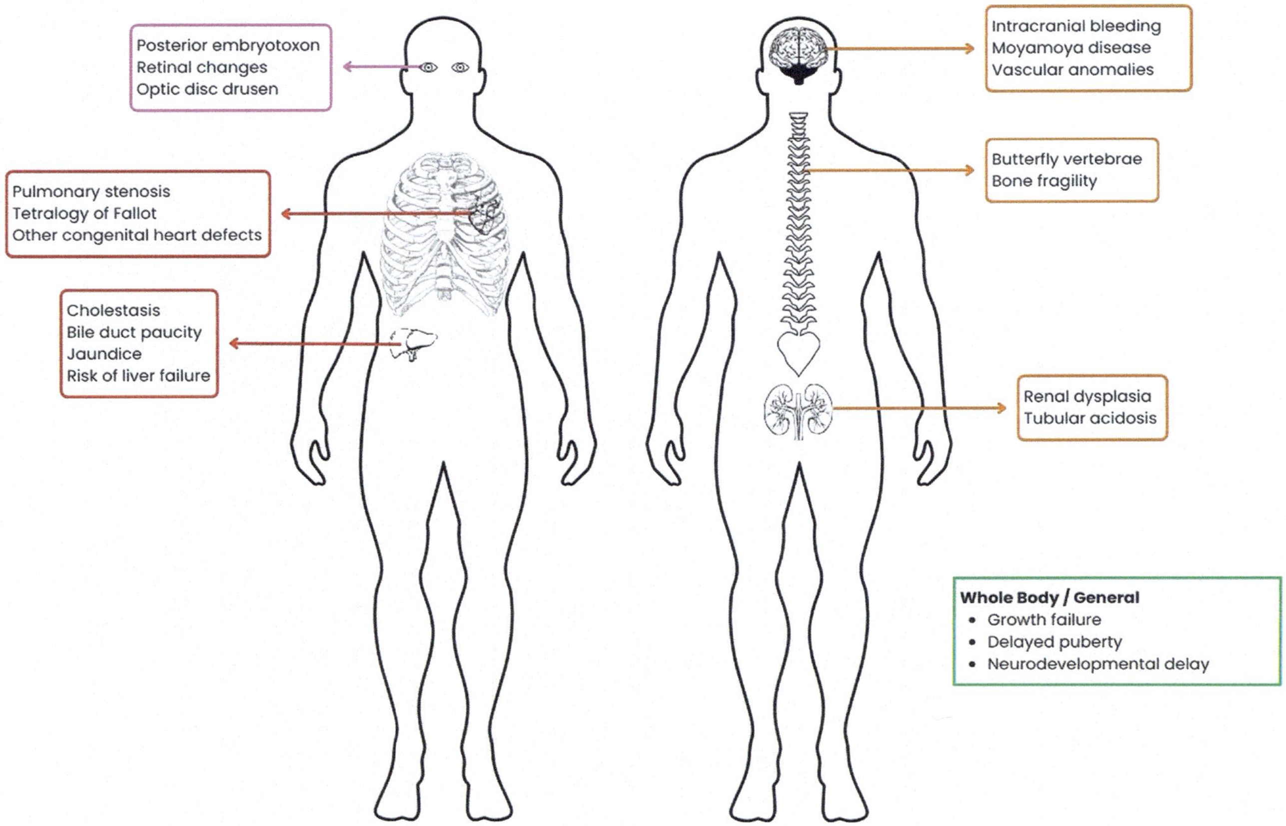

FIGURE 8.2

Systemic manifestations of Alagille syndrome. An annotated body diagram illustrating the major organ systems affected, including hepatic, cardiac, ocular, neurological, skeletal, and renal involvement, as well as general growth and developmental features. *ALGS*, Alagille syndrome.

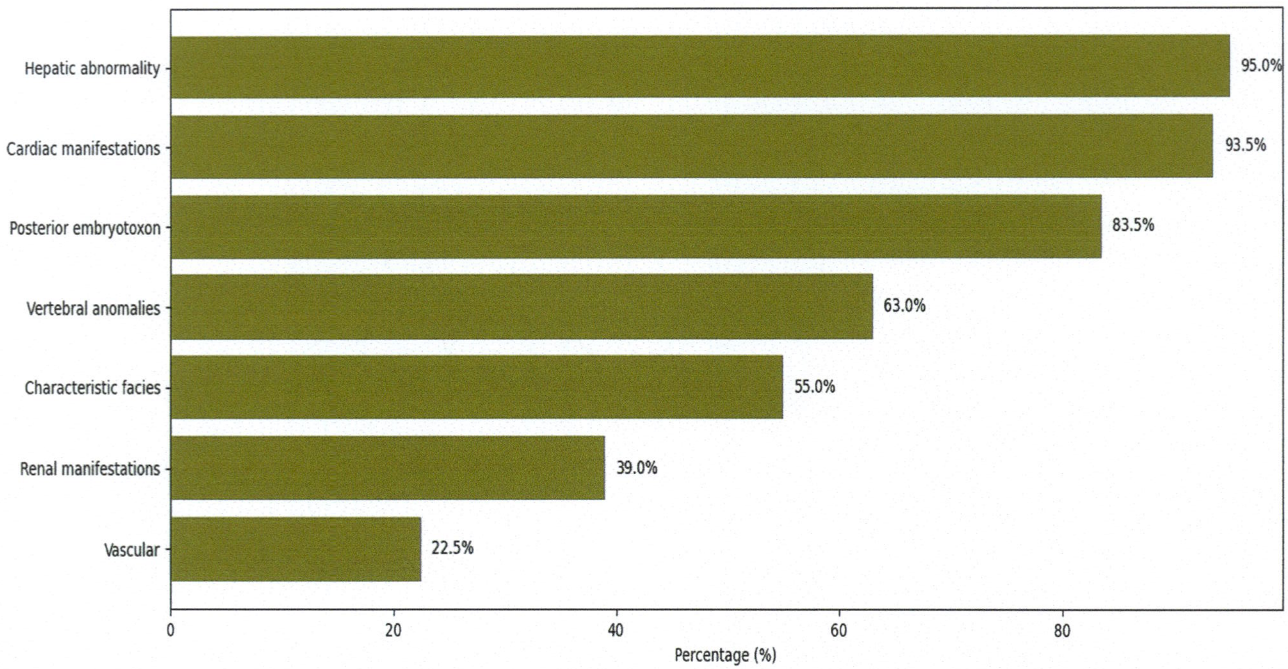

FIGURE 8.3

Prevalence of major clinical features in Alagille syndrome. Bar graph showing the reported frequency of seven key manifestations.

abnormalities such as posterior embryotoxon can be seen in 51% of patients. Peripheral chorioretinal changes and optic nerve drusen can also be observed. Formation of butterfly vertebrae is seen due to the failure of fusion of anterior arches. Radio-ulnar synostosis and rib anomalies may occur due to disruption in Notch signaling. The kidneys are affected in around 38% of the patients. Congenital anomalies of the kidney and urinary tract (CAKUT), such as renal hypoplasia, dysplasia, cystic changes, and urinary tract obstruction, are among the most common patterns. Functional abnormalities include renal tubular acidosis, renovascular disease, decreased glomerular filtration rate, and various forms of glomerulopathy, including glomerular mesangiolipidosis and proteinuria. Experimental and human data suggest that aberrant Notch1 activation in glomerular endothelial cells may impair expression of VE-cadherin and promote albuminuria via transcriptional regulators such as Snail and Erg. Despite this spectrum of kidney disease, renal transplantation remains relatively uncommon compared with liver transplantation (Gilbert et al., 2019). Vascular abnormalities of the central nervous system and systemic circulation are increasingly recognized and contribute to neurologic risk. Intracranial bleeding, reported in a notable subset of patients, may result from aneurysms, moyamoya-like arteriopathy, or stenosis and dysplasia of the internal carotid, middle cerebral, and basilar arteries. These vascular lesions put patients at risk of both ischemic and hemorrhagic strokes, including subdural, epidural, and subarachnoid haemorrhages, which may occur after only minor trauma or even spontaneously. In addition, systemic arterial involvement, such as aortic coarctation and other large-vessel anomalies, has been reported, highlighting the widespread vasculopathy that results from disrupted Notch signaling in ALGS (Turnpenny & Ellard, 2012). Overall, the clinical picture of ALGS mirrors the wide-ranging role of the Notch pathway in early development: the same underlying defect can affect the liver, heart, face, eyes, bones, kidneys and blood vessels in very different ways, and the severity and combination of these problems can vary markedly even between family members who carry the identical pathogenic variant (Ahn et al., 2015).

3.2 Hypophosphatasia

HPP is an inherited bone disorder in which loss-of-function variants in the ALPL gene reduce or abolish the activity of TNSALP, the key alkaline phosphatase isoenzyme in bone, liver, and kidney. More than 400 ALPL mutations have been identified, encompassing missense, nonsense, splice-site variants, and small or large deletions, and giving rise to a broad phenotypic continuum. TNSALP is a homodimeric enzyme with multiple metal-binding sites (two zinc, one magnesium, and an additional calcium site not present in bacterial alkaline phosphatase) and dephosphorylates key extracellular substrates, including inorganic pyrophosphate (PPi), pyridoxal-5′-phosphate (PLP), and phosphoethanolamine (PEA) (Im & Cho, 2025). When TNSALP activity is reduced, PPi accumulates in the extracellular matrix, disrupting the balance between phosphate and PPi and inhibiting hydroxyapatite crystal formation in bone and teeth. Clinically, this leads to rickets or osteomalacia-like mineralization defects, recurrent fractures, bone pain, chronic musculoskeletal disability, and premature loss of deciduous teeth, often with intact tooth roots (Fig. 8.4). Chondrocalcinosis from calcium pyrophosphate crystal deposition and pseudogout-like acute arthritis are well-recognized complications in older children and adults. Hypercalcemia and hypercalciuria can occur, predisposing to nephrocalcinosis and progressive decline in renal function, sometimes accompanied by suppression of parathyroid hormone (Im & Cho, 2025). PLP, the active form of vitamin B6, is normally dephosphorylated by TNSALP at the blood–brain barrier to pyridoxal, which can cross into the central nervous system and be rephosphorylated back to PLP. In severe HPP, impaired PLP handling can produce functional vitamin B6 deficiency in the brain, leading to pyridoxine-responsive seizures in infancy. Experimental work also suggests that TNSALP influences neural stem cell proliferation, myelination, axonal growth, and synapse maturation, which may contribute to the neurological phenotype in some patients (Godinez et al., 2022). Clinically, HPP is classified into six principal forms according to age at onset and

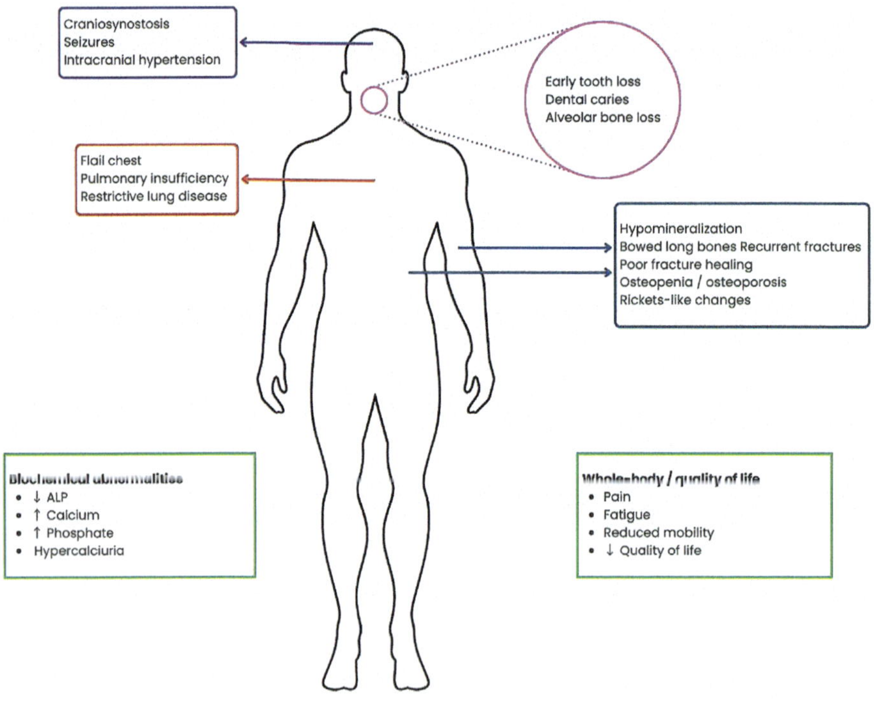

FIGURE 8.4

Systemic manifestations of Hypophosphatasia. Annotated body diagram depicting the multisystem features of Hypophosphatasia, including neurological, respiratory, dental, and skeletal complications, alongside characteristic biochemical abnormalities and their impact on quality of life. *ALP*, alkaline phosphatase.

predominant manifestations: perinatal lethal, perinatal benign, infantile, childhood, adult, and odontohypophosphatasia. Perinatal lethal HPP manifests in utero or at birth with profound skeletal hypomineralization, shortened and bowed long bones, skin-covered osteochondral spurs, respiratory failure from hypoplastic lungs, and high perinatal mortality; survivors have severe rachitic chest deformities, apnea, and seizures. Perinatal benign HPP can look very worrying on late-pregnancy scans, but bone mineralization often improves on its own in the third trimester or early infancy, so many babies do far better than initially expected. By contrast, infantile and childhood HPP usually show up after birth with poor growth, soft or painful bones, frequent fractures, and early loss of baby teeth, with severity ranging from life-threatening disease in infants to milder bone and dental problems in older children (Pattie & Iffland, 2025). Symptoms may partially remit during adolescence, but skeletal disease can re-emerge later in life. Adult HPP usually presents in middle age with stress fractures or pseudofractures of the metatarsals and femora, chronic bone and joint pain, muscle weakness, and increased risk of chondrocalcinosis and osteoarthropathy. A history of childhood skeletal symptoms or premature loss of primary teeth is often elicited retrospectively. Misdiagnosis as osteoporosis is common, and treatment with bisphosphonates, which further inhibit bone mineralization, can exacerbate skeletal fragility and promote atypical femoral fractures in these patients. Adult HPP usually presents in middle age with stress fractures or pseudofractures of the metatarsals and femora, chronic bone and joint pain, muscle weakness, and increased risk of chondrocalcinosis and osteoarthropathy. A history of childhood skeletal symptoms or premature loss of primary teeth is often elicited retrospectively. Misdiagnosis as osteoporosis is common, and treatment with bisphosphonates, which further inhibit bone mineralization, can exacerbate skeletal fragility and promote atypical femoral fractures in these patients. Another manifestation found in HPP is odontohypophosphatasia (Whyte, 2017). This anomaly is characterized predominantly by dental manifestations in the absence of overt skeletal disease. It can appear in children or adults and often first shows up as premature loss of fully rooted baby teeth, especially the front incisors, sometimes with prominent dental caries. X-rays may show reduced alveolar bone and enlarged pulp chambers and root canals, and anyone with unexplained early tooth loss should be evaluated for HPP more broadly.

4. Diagnostic approaches

ALGS is usually suspected when a child shows a combination of problems in several organs. Most often, the liver and heart, together with typical facial features and sometimes the kidney, eye, bone, or blood vessel, are abnormally developed (Gaio et al., 2021, pp. 57–83). The diagnosis is then confirmed by finding a heterozygous pathogenic variant in JAG1 (in most cases) or, less often, NOTCH2. In practice, evaluation often starts in an infant with cholestasis, where blood tests show conjugated hyperbilirubinaemia and raised bile acids, cholesterol, GGT, and transaminases. A liver biopsy is now mainly used to distinguish ALGS from biliary atresia or other cholangiopathies; reduced numbers of interlobular bile ducts can support ALGS but are not specific and must be interpreted alongside the clinical picture and genetics. Once ALGS is on the radar, a structured "head-to-toe" work-up is essential, including echocardiography for pulmonary artery stenosis or other heart defects, spine X-rays for butterfly vertebrae, slit-lamp examination for posterior embryotoxon, and renal ultrasound with Doppler to look for congenital kidney and renovascular anomalies. Abdominal ultrasonography may show a small or absent gallbladder in a subset of patients, but this is not specific and overlaps with biliary atresia (Sergi et al., 2008). Genetic confirmation by JAG1/NOTCH2 analysis both secures the diagnosis and enables cascade testing in relatives and consideration of prenatal or preimplantation diagnosis, although phenotypic severity cannot be predicted from genotype alone (Gilbert et al., 2019; Leonard et al., 2014; Saleh et al., 2016; Sharma et al., 2026; Turnpenny & Ellard, 2012).

HPP is suspected in individuals with persistently and clearly reduced serum alkaline phosphatase (ALP) activity for age and sex, particularly when accompanied by skeletal, dental, or mineralization abnormalities. Secondary causes of low ALP, such as antiresorptive therapy (bisphosphonates, denosumab), excessive vitamin D, recent chemotherapy, untreated hypothyroidism, severe anemia, pregnancy-related changes, and hypomagnesemia, must first be excluded. Elevated plasma pyridoxal-5′-phosphate (PLP) and sometimes increased urinary phosphoethanolamine (PEA) provide supportive biochemical evidence, although PEA is not specific and may be normal in some patients (Rush et al., 2024; Shajani-Yi et al., 2020). Radiographs help define the extent and pattern of bone disease. In infants and children, typical findings include generalized osteopenia, rickets-like metaphyseal changes, craniosynostosis, and poorly mineralized long bones. In adolescents and adults, pseudofractures, stress fractures of the metatarsals or femur, and chondrocalcinosis are common, often in the setting of chronic musculoskeletal pain (Im & Cho, 2025). Definitive diagnosis relies on identifying pathogenic or likely pathogenic variants in ALPL, which encodes tissue-nonspecific alkaline phosphatase, typically by sequencing of the coding regions and exon–intron boundaries, with copy-number analysis when indicated, and detects the majority of causative variants, though some lie in deep intronic or regulatory regions. Current guidelines emphasize that HPP should be diagnosed by integrating low ALP, elevated PLP (and/or PEA), compatible clinical and radiographic features, and ALPL genotype, rather than relying on any single parameter in isolation. In at-risk pregnancies where parental ALPL variants are known, prenatal diagnosis can be performed via chorionic villus sampling or amniocentesis, supported by fetal ultrasound findings of skeletal hypomineralization, although sonographic changes are not specific for HPP (Kishnani et al., 2025; Michigami et al., 2020; Rush et al., 2024).

5. NIPS for Alagille syndrome

NIPS for ALGS is now technically possible using single-gene NIPT panels. These panels examine tiny fragments of fetal DNA with next-generation sequencing

(NGS). These assays target coding exons and exon–intron boundaries in a defined set of genes associated with autosomal dominant or de novo single-gene disorders, including JAG1 and other ALGS-related genes. The largest published clinical experience comes from Mohan et al. (2022), who evaluated an NIPT panel (commercially offered as Vistara) covering 30 genes and 25 dominant conditions, among them ALGS, in 2208 singleton pregnancies from 9 weeks' gestation onwards. Testing involved parallel sequencing of cffDNA from maternal plasma together with maternal and paternal genomic DNA (trio analysis) to improve discrimination between fetal and parental variants and to support accurate variant interpretation. Overall, 5.7% of women screened positive for a single-gene disorder, and in the subset referred for ultrasound-detected long-bone abnormalities, one positive result corresponded to ALGS. Among cases with available confirmatory testing, no false-positive or false-negative results were observed, demonstrating high analytical performance in a high-risk cohort (Deans et al., 2013; Shea et al., 2013).

From a clinical standpoint, NIPS for ALGS is most informative in pregnancies with a known familial JAG1/NOTCH2 pathogenic variant or with ultrasound findings suggestive of a disorder on the panel (e.g., major cardiac defects, characteristic craniofacial or skeletal anomalies) rather than as a general population screen. A positive NIPT result for an ALGS-associated variant remains a screening outcome and must be verified by invasive diagnostic testing, chorionic villus sampling, or amniocentesis with targeted sequencing, before any irreversible decisions are taken. Likewise, a negative result does not completely exclude ALGS, because current panels cover only known pathogenic variants in predefined regions and may miss rare or novel changes, deep intronic variants, or low fetal fractions. Current guidelines therefore view single-gene NIPT for ALGS as an adjunct to, not a replacement for, established diagnostic pathways: it can provide earlier, risk-free information in high-risk pregnancies, help prioritize invasive testing, and facilitate multidisciplinary counseling, but confirmatory diagnostic procedures remain mandatory for definitive prenatal diagnosis and pregnancy management (Chen et al., 2025; Mohan et al., 2022; Shea et al., 2013).

6. NIPS for hypophosphatasia

By contrast, NIPS for HPP is still mostly experimental, and, in practice, prenatal diagnosis still depends on invasive genetic testing together with detailed ultrasound assessment of the fetal skeleton. For families with a known ALPL pathogenic variant, standard practice is targeted molecular testing of chorionic villus or amniotic fluid DNA, sometimes complemented by measurement of tissue-nonspecific alkaline phosphatase activity in cultured amniocytes or chorionic villus cells and careful fetal skeletal assessment by ultrasound. Classic studies have shown that absent or severely reduced mineralization of the calvarium and long bones, a narrow thorax, and beaded ribs on mid-trimester ultrasound, together with low ALP isoenzyme activity in cultured amniotic cells, provide robust prenatal evidence for severe, early-onset HPP.

At a technical level, the same cffDNA-based approaches used for other monogenic diseases could, in principle, be adapted to detect known familial ALPL variants non-invasively, for example, by targeted NGS or digital PCR on maternal plasma, sometimes in combination with parental genotyping and relative haplotype dosage analysis. However, to date, there are no large clinical series validating single-gene NIPT specifically for HPP, and its use remains confined to research or highly specialized settings. Key challenges include the broad allelic heterogeneity of ALPL, the need to distinguish maternal carrier status from fetal genotype, and the requirement for very high analytical sensitivity and specificity given the severe implications of a prenatal diagnosis (Kishnani et al., 2025; Michigami et al., 2020). For now, therefore, prenatal evaluation of suspected HPP is best approached through a combination of:

- Detailed ultrasound surveillance for skeletal hypomineralization and thoracic deformity.
- Invasive molecular testing for known parental ALPL variants (or broader sequencing when the familial variant is unknown).
- Biochemical studies on fetal or cultured cells were available (Mulivor et al., 1978).

Non-invasive approaches using cffDNA should be regarded as experimental and, if offered, must be embedded within research protocols or highly specialized services with access to confirmatory invasive testing, robust variant interpretation, and comprehensive pre- and post-test genetic counseling (Deans et al., 2013; Li et al., 2009; Sankaran et al., 2020).

7. Comparison with traditional diagnostic methods

NIPS offers a range of advantages over traditional diagnostic methods. Although, they need to be supplemented with traditional diagnostic methods currently, they have had fewer false positives (comparatively) and even fewer false negatives. A meta-analysis of NIPS cases revealed this FN rate to be 0.0079% in the cohort (Liehr, 2022). NIPS uses the cell-free fetal DNA, which can be identified as early as around the 10th week of gestation, when the median cfDNA is 10.2%, leading to earlier detection than traditional diagnostics (Wang et al., 2013). This provides the early psychological relief that mothers need during this time and helps them make an informed decision. In case of adverse genetic conditions, NIPS can help decide on more monitoring and even early pregnancy termination. NIPS reduces the risk of prenatal testing miscarriages. A study found this to be 0.11% for amniocentesis and 0.22% for CVS (Akolekar et al., 2015).

8. Ethical and legal considerations

NIPS can improve access to prenatal genetics, but its ethical and legal acceptability depends on how it is funded, framed, and regulated. NIPS are an alternative mode of screening for people in the lower socio-economic strata (Kater-Kuipers et al., 2018). These provide them with the tools to make decisions, but in a much more economical way.

It is, however, imperative that these NIPS are offered in conjunction with genetic counseling. This is to ensure false positives are dealt with and to help patients make a better informed decision. Although large meta-analyses show that NIPS has very low false-negative rates, false positives are more frequent, especially for rare conditions and microdeletions, and any high-risk result still requires confirmatory invasive testing (Liehr, 2022; Perrot & Horn, 2022; Zaami et al., 2021). Many policy and ethics reports suggest that NIPS should be used mainly for serious, early-onset conditions, not for non-medical purposes like routine sex selection. They also emphasize that counseling should cover not only medical risks but also the support, treatment options, and possibilities for a good quality of life for people living with these genetic conditions (Perrot & Horn, 2022; Raz, 2005).

9. Conclusion

ALGS and HPP show how a single gene change can produce complex, whole-body disease. In ALGS, JAG1 (and sometimes NOTCH2) variants disturb Notch signaling. It manifests with system-wide disruptions spanning the hepatic, cardiac, renal, skeletal, ophthalmologic, and vascular systems. In HPP, the loss of TNSALP activity produces consequences extending beyond the skeleton, disrupts vitamin B6 metabolism, impairs CNS function, and contributes to chronic pain and organ damage across the lifespan. NIPS utilizes the cfDNA present in the maternal plasma. Using whole genome sequencing techniques, lethal mutations in genes can be discovered and make decisions regarding termination of pregnancy, to prevent an infant from being born with severe abnormalities.

References

Ahn, K. J., Yoon, J. K., Kim, G. B., et al. (2015). Alagille syndrome and a JAG1 mutation: 41 cases of experience at a single center. *Korean Journal of Pediatrics, 58*(10), 392–397. https://doi.org/10.3345/kjp.2015.58.10.392

Akolekar, R., Beta, J., Picciarelli, G., Ogilvie, C., & D'Antonio, F. (2015). Procedure-related risk of miscarriage following amniocentesis and chorionic villus sampling: A systematic review and meta-analysis. *Ultrasound in Obstet & Gyne, 45*(1), 16–26. https://doi.org/10.1002/uog.14636

Alagille, D., Odièvre, M., Gautier, M., & Dommergues, J. P. (1975). Hepatic ductular hypoplasia associated with characteristic facies, vertebral malformations, retarded physical, mental, and sexual development, and cardiac murmur. *The Journal of pediatrics, 86*(1), 63–71. https://doi.org/10.1016/s0022-3476(75)80706-2

American College of Obstetricians and Gynecologists Committee on Practice Bulletins—Obstetrics Committee on Genetics Society for Maternal-Fetal Medicine. (2020).

Screening for fetal chromosomal abnormalities: ACOG practice bulletin, number 226. *Obstetrics & Gynecology, 136*(4), e48–e69. https://doi.org/10.1097/AOG.0000000000004084

Ashoor, G., Syngelaki, A., Poon, L. C. Y., Rezende, J. C., & Nicolaides, K. H. (2013). Fetal fraction in maternal plasma cell-free DNA at 11–13 weeks' gestation: Relation to maternal and fetal characteristics. *Ultrasound in Obstet & Gyne, 41*(1), 26–32. https://doi.org/10.1002/uog.12331

Ayoub, M. D., & Kamath, B. M. (2020). Alagille syndrome: Diagnostic challenges and advances in management. *Diagnostics, 10*(11), 907. https://doi.org/10.3390/diagnostics10110907

Ayoub, M. D., & Kamath, B. M. (2022). Alagille syndrome. *Clinics in Liver Disease, 26*(3), 355–370. https://doi.org/10.1016/j.cld.2022.03.002

Chen, S., Xu, W., Zhang, L., et al. (2025). Expanded noninvasive prenatal screening for dominant single-gene disorders: Proof-of-concept, performance, and challenges. *American Journal of Obstetrics and Gynecology, 233*(6), 673.e1–673.e21. https://doi.org/10.1016/j.ajog.2025.06.043

Da Palma, M. M., Igelman, A. D., Ku, C., et al. (2021). Characterization of the spectrum of ophthalmic changes in patients with alagille syndrome. *Investigative Ophthalmology & Visual Science, 62*(7), 27. https://doi.org/10.1167/iovs.62.7.27

Dahir, K. M., Seefried, L., Kishnani, P. S., et al. (2022). Clinical profiles of treated and untreated adults with hypophosphatasia in the Global HPP Registry. *Orphanet Journal of Rare Diseases, 17*(1), 277. https://doi.org/10.1186/s13023-022-02393-8

Deans, Z., Hill, M., Chitty, L. S., & Lewis, C. (2013). Non-invasive prenatal testing for single gene disorders: Exploring the ethics. *European Journal of Human Genetics, 21*(7), 713–718. https://doi.org/10.1038/ejhg.2012.250

Duffus, S., Thrasher, B., & Calikoglu, A. S. (2018). Brief clinical report: Hypophosphatasia—Diagnostic considerations and treatment outcomes in an infant. In Y. J. Lee (Ed.), *Case reports in pediatrics* (Vol 2018), Article 5719761. https://doi.org/10.1155/2018/5719761

El-Koofy, N. M., El-Mahdy, R., Fahmy, M. E., El-Hennawy, A., Farag, M. Y., & El-Karaksy, H. M. (2011). Alagille syndrome: Clinical and ocular pathognomonic features. *European Journal of Ophthalmology, 21*(2), 199–206. https://doi.org/10.5301/EJO.2010.5675

Emerick, K. M., Krantz, I. D., Kamath, B. M., et al. (2005). Intracranial vascular abnormalities in patients with alagille syndrome. *Journal of Pediatric Gastroenterology and Nutrition, 41*(1), 99–107. https://doi.org/10.1097/01.MPG.0000162776.67758.2F

Emerick, K. M., Rand, E. B., Goldmuntz, E., Krantz, I. D., Spinner, N. B., & Piccoli, D. A. (1999). Features of alagille syndrome in 92 patients: Frequency and relation to prognosis. *Hepatology, 29*(3), 822–829. https://doi.org/10.1002/hep.510290331

Fauvert, D., Brun-Heath, I., Lia-Baldini, A. S., et al. (2009). Mild forms of hypophosphatasia mostly result from dominant negative effect of severe alleles or from compound heterozygosity for severe and moderate alleles. *BMC Medical Genetics, 10*(1), 51. https://doi.org/10.1186/1471-2350-10-51

Gaio, P., Reffo, E., Mescoli, C., & Cananzi, M. (2021). *Alagille syndrome. In diseases of the liver and biliary tree*. Cham: Springer International Publishing.

Gilbert, M. A., Bauer, R. C., Rajagopalan, R., et al. (2019). Alagille syndrome mutation update: Comprehensive overview of JAG1 and NOTCH2 mutation frequencies and insight into missense variant classification. *Human Mutation, 40*(12), 2197–2220. https://doi.org/10.1002/humu.23879

Godinez, A., Rajput, R., Chitranshi, N., Gupta, V., Basavarajappa, D., Sharma, S., You, Y., Pushpitha, K., Dhiman, K., Mirzaei, M., & Graham, S. (2022). Neuroserpin, a crucial regulator for axogenesis, synaptic modelling and cell–cell interactions in the pathophysiology of neurological disease. *Cellular and Molecular Life Sciences, 79*(3), 172.

Hanson, B., Scotchman, E., Chitty, L. S., & Chandler, N. J. (2022). Non-invasive prenatal diagnosis (NIPD): How analysis of cell-free DNA in maternal plasma has changed prenatal diagnosis for monogenic disorders. *Clinical Science, 136*(22), 1615–1629. https://doi.org/10.1042/CS20210380

Hingorani, M., Nischal, K. K., Davies, A., et al. (1999). Ocular abnormalities in Alagille syndrome11None of the authors has any commercial interest arising from the findings presented in this article. *Ophthalmology, 106*(2), 330–337. https://doi.org/10.1016/S0161-6420(99)90072-6

Huggins, E., Ong, R., Rockman-Greenberg, C., Flueckinger, L. B., Dahir, K. M., & Kishnani, P. S. (2020). Multigenerational case examples of hypophosphatasia: Challenges in genetic counseling and disease management. *Molecular Genetics and Metabolism Reports, 25*, Article 100661. https://doi.org/10.1016/j.ymgmr.2020.100661

Im, M., & Cho, S. Y. (2025). Hypophosphatasia in childhood: Diagnosis to management. *Osteoporos Sarcopenia, 11*(2), 38–42. https://doi.org/10.1016/j.afos.2025.05.003

Kamath, B. M., Bason, L., Piccoli, D. A., Krantz, I. D., & Spinner, N. B. (2003). Consequences of JAG1 mutations. *Journal of Medical Genetics, 40*(12), 891–895. https://doi.org/10.1136/jmg.40.12.891

Kamath, B. M., Chen, Z., Romero, R., et al. (2015). Quality of life and its determinants in a multicenter cohort of children with alagille syndrome. *The Journal of Pediatrics, 167*(2), 390–396.e3. https://doi.org/10.1016/j.jpeds.2015.04.077

Kamath, B. M., Spinner, N. B., Emerick, K. M., et al. (2004). Vascular anomalies in alagille syndrome: A significant cause of morbidity and mortality. *Circulation, 109*(11), 1354–1358. https://doi.org/10.1161/01.CIR.0000121361.01862.A4

Kamath, B. M., Stolle, C., Bason, L., et al. (2002). Craniosynostosis in Alagille syndrome. *American Journal of Medical Genetics, 112*(2), 176–180. https://doi.org/10.1002/ajmg.10608

Kater-Kuipers, A., Bunnik, E. M., De Beaufort, I. D., & Galjaard, R. J. H. (2018). Limits to the scope of non-invasive prenatal testing (NIPT): An analysis of the international ethical framework for prenatal screening and an interview study with Dutch professionals. *BMC Pregnancy and Childbirth, 18*(1), 409. https://doi.org/10.1186/s12884-018-2050-4

Khan, A. A., Brandi, M. L., Rush, E. T., et al. (2024). Hypophosphatasia diagnosis: Current state of the art and proposed diagnostic criteria for children and adults. *Osteoporosis International, 35*(3), 431–438. https://doi.org/10.1007/s00198-023-06844-1

Kishnani, P. S., Rehder, C., Ozono, K., et al. (2025). Revisiting the genetics of hypophosphatasia. *Journal of Inherited Metabolic Disease, 48*(6), Article e70083. https://doi.org/10.1002/jimd.70083

Kohut, T. J., Gilbert, M. A., & Loomes, K. M. (2021). Alagille syndrome: A focused review on clinical features, genetics, and treatment. *Seminars in Liver Disease, 41*(04), 525–537. https://doi.org/10.1055/s-0041-1730951

Krantz, I. D., Piccoli, D. A., & Spinner, N. B. (1997). Alagille syndrome. *Journal of Medical Genetics, 34*(2), 152–157. https://doi.org/10.1136/jmg.34.2.152

Leonard, L. D., Chao, G., Baker, A., Loomes, K., & Spinner, N. B. (2014). Clinical utility gene card for: Alagille Syndrome (ALGS). *European Journal of Human Genetics, 22*(3), 435. https://doi.org/10.1038/ejhg.2013.140

Leung, D. H., Sorensen, L. G., Ye, W., et al. (2022). Neurodevelopmental outcomes in children with inherited liver disease and native liver. *Journal of Pediatric Gastroenterology and Nutrition, 74*(1), 96–103. https://doi.org/10.1097/MPG.0000000000003337

Li, Y., Altarescu, G., Renbaum, P., et al. (2009). Non-invasive prenatal diagnosis using cell-free fetal DNA in maternal plasma from PGD pregnancies. *Reproductive BioMedicine Online, 19*(5), 714–720. https://doi.org/10.1016/j.rbmo.2009.09.005

Liehr, T. (2022). False-positives and false-negatives in non-invasive prenatal testing (NIPT): What can we learn from a meta-analyses on > 750,000 tests? *Molecular Cytogenetics, 15*(1), 36. https://doi.org/10.1186/s13039-022-00612-2

Lin, H. C., Le Hoang, P., Hutchinson, A., et al. (2012). Alagille syndrome in a Vietnamese cohort: Mutation analysis and assessment of facial features. *American Journal of Medical Genetics—Part A, 158A*(5), 1005–1013. https://doi.org/10.1002/ajmg.a.35255

Lira Dos Santos, E. J., Mohamed, F. F., Kramer, K., & Foster, B. L. (2025). Dental manifestations of hypophosphatasia: Translational and clinical advances. *JBMR Plus, 9*(2), ziae180. https://doi.org/10.1093/jbmrpl/ziae180

Lo, Y. M. D., Corbetta, N., Chamberlain, P. F., et al. (1997). Presence of fetal DNA in maternal plasma and serum. *The Lancet, 350*(9076), 485–487. https://doi.org/10.1016/S0140-6736(97)02174-0

Loomes, K. M., Spino, C., Goodrich, N. P., et al. (2019). Bone density in children with chronic liver disease correlates with growth and cholestasis. *Hepatology, 69*(1), 245–257. https://doi.org/10.1002/hep.30196

McElhinney, D. B., Krantz, I. D., Bason, L., et al. (2002). Analysis of cardiovascular phenotype and genotype-phenotype correlation in individuals with a JAG1 mutation and/or Alagille syndrome. *Circulation, 106*(20), 2567–2574. https://doi.org/10.1161/01.CIR.0000037221.45902.69

Michigami, T., Ohata, Y., Fujiwara, M., et al. (2020). Clinical practice guidelines for hypophosphatasia. *Clinical Pediatric Endocrinology, 29*(1), 9–24. https://doi.org/10.1297/cpe.29.9

Mitchell, E., Gilbert, M., & Loomes, K. M. (2018). Alagille syndrome. *Clinics in Liver Disease, 22*(4), 625–641. https://doi.org/10.1016/j.cld.2018.06.001

Mohan, P., Lemoine, J., Trotter, C., et al. (2022). Clinical experience with non-invasive prenatal screening for single-gene disorders. *Ultrasound in Obstet & Gyne, 59*(1), 33–39. https://doi.org/10.1002/uog.23756

Moliner, A. M., & Waligora, J. (2017). The European Union Policy in the field of rare diseases. *Advances in Experimental Medicine and Biology, 1031*, 561–587. https://doi.org/10.1007/978-3-319-67144-4_30

Mori, M., DeArmey, S. L., Weber, T. J., & Kishnani, P. S. (2016). Case series: Odontohypophosphatasia or missed diagnosis of childhood/adult-onset hypophosphatasia?—Call for a long-term follow-up of premature loss of primary teeth. *Bone Reports, 5*, 228–232. https://doi.org/10.1016/j.bonr.2016.08.004

Mornet, E. (2007). Hypophosphatasia. *Orphanet Journal of Rare Diseases, 2*(1), 40. https://doi.org/10.1186/1750-1172-2-40

Mornet, E. (2008). Hypophosphatasia. *Best Practice & Research Clinical Rheumatology, 22*(1), 113–127. https://doi.org/10.1016/j.berh.2007.11.003

Mornet, E. (2017). Genetics of hypophosphatasia. *Archives de Pediatrie, 24*(5), 5S51–5S56. https://doi.org/10.1016/S0929-693X(18)30014-9

Mornet, E., Taillandier, A., Domingues, C., et al. (2021). Hypophosphatasia: A genetic-based nosology and new insights in genotype-phenotype correlation. *European Journal of Human Genetics, 29*(2), 289–299. https://doi.org/10.1038/s41431-020-00732-6

Mornet, E., Yvard, A., Taillandier, A., Fauvert, D., & Simon-Bouy, B. (2011). A molecular-based estimation of the prevalence of hypophosphatasia in the European population. *Annals of Human Genetics, 75*(3), 439–445. https://doi.org/10.1111/j.1469-1809.2011.00642.x

Mouzaki, M., Bass, L. M., Sokol, R. J., et al. (2016). Early life predictive markers of liver disease outcome in an International, Multicentre Cohort of children with Alagille syndrome. *Liver International, 36*(5), 755–760. https://doi.org/10.1111/liv.12920

Mulivor, R. A., Mennuti, M., Zackai, E. H., & Harris, H. (1978). Prenatal diagnosis of hypophosphatasia; genetic, biochemical, and clinical studies. *The American Journal of Human Genetics, 30*(3), 271–282.

Nguengang Wakap, S., Lambert, D. M., Olry, A., et al. (2020). Estimating cumulative point prevalence of rare diseases: Analysis of the Orphanet database. *European Journal of Human Genetics, 28*(2), 165–173. https://doi.org/10.1038/s41431-019-0508-0

Paez-Escamilla, M., Scanga, H. L., Liasis, A., & Nischal, K. K. (2022). Macular atrophy in JAG1- related Alagille syndrome: A case series. *Ophthalmic Genetics, 43*(2), 230–234. https://doi.org/10.1080/13816810.2021.2004432

Pattie, E. A., & Iffland, P. H. (2025). Shared disease mechanisms in neurodevelopmental disorders: A cellular and molecular biology perspective. *Brain Sciences, 16*(1), 54.

Perrot, A., & Horn, R. (2022). The ethical landscape(s) of non-invasive prenatal testing in England, France and Germany: Findings from a comparative literature review. *European Journal of Human Genetics, 30*(6), 676–681. https://doi.org/10.1038/s41431-021-00970-2

Pierpont, E. I., Simmons, J. H., Spurlock, K. J., Shanley, R., & Sarafoglou, K. M. (2021). Impact of pediatric hypophosphatasia on behavioral health and quality of life. *Orphanet Journal of Rare Diseases, 16*(1), 80. https://doi.org/10.1186/s13023-021-01722-7

Raz, A. E. (2005). Disability rights, prenatal diagnosis and eugenics: A cross-cultural view. *Journal of Genetic Counseling, 14*(3), 183–187. https://doi.org/10.1007/s10897-005-0573-0

Reynal, F., Camoin, A., Tardieu, C., Fabre, A., & Blanchet, I. (2023). Oral findings in children with congenital cholestatic disease: A systematic review of case reports and case series. *Archives de Pediatrie, 30*(6), 427–437. https://doi.org/10.1016/j.arcped.2023.06.003

Rocha, R., Soro, I., Leitão, A., Silva, M. L., & Leão, M. (2012). Moyamoya vascular pattern in Alagille syndrome. *Pediatric Neurology, 47*(2), 125–128. https://doi.org/10.1016/j.pediatrneurol.2012.04.014

Rodrigues Bento, J., Krebsová, A., Van Gucht, I., et al. (2022). Isolated aneurysmal disease as an underestimated finding in individuals with JAG1 pathogenic variants. *Human Mutation, 43*(12), 1824–1828. https://doi.org/10.1002/humu.24433

Rush, E., Brandi, M. L., Khan, A., et al. (2024). Proposed diagnostic criteria for the diagnosis of hypophosphatasia in children and adolescents: Results from the HPP International Working Group. *Osteoporosis International, 35*(1), 1–10. https://doi.org/10.1007/s00198-023-06843-2

Saleh, M., Kamath, B. M., & Chitayat, D. (2016). Alagille syndrome: Clinical perspectives. *The Application of Clinical Genetics, 9*, 75–82. https://doi.org/10.2147/TACG.S86420

Salem, J. E., Bruguiere, E., Iserin, L., Guiochon-Mantel, A., & Plouin, P. F. (2012). Hypertension and aortorenal disease in Alagille syndrome. *Journal of Hypertension, 30*(7), 1300–1306. https://doi.org/10.1097/HJH.0b013e3283531e1f

Sankaran, D., Chandrasekharan, P. K., & Rawat, M. (2020). Perinatal hypophosphatasia in a premature infant. *American Journal of Perinatology Reports, 10*(2), e139–e147. https://doi.org/10.1055/s-0040-1709512

Schindler, E. A., Gilbert, M. A., Piccoli, D. A., Spinner, N. B., Krantz, I. D., & Loomes, K. M. (2021). Alagille syndrome and risk for hepatocellular carcinoma: Need for increased surveillance in adults with mild liver phenotypes. *American Journal of Medical Genetics—Part A, 185*(3), 719–731. https://doi.org/10.1002/ajmg.a.62028

Schroth, R. J., Long, C., Lee, V. H. K., Alai-Towfigh, H., & Rockman-Greenberg, C. (2021). Dental outcomes for children receiving asfotase alfa for hypophosphatasia. *Bone, 152*, Article 116089. https://doi.org/10.1016/j.bone.2021.116089

Seefried, L., Dahir, K., Petryk, A., et al. (2020). Burden of illness in adults with hypophosphatasia: Data from the global hypophosphatasia patient registry. *Journal of Bone and Mineral Research, 35*(11), 2171–2178. https://doi.org/10.1002/jbmr.4130

Seefried, L., Genest, F., Hofmann, C., Brandi, M. L., & Rush, E. (2025). Diagnosis and treatment of hypophosphatasia. *Calcified Tissue International, 116*(1), 46. https://doi.org/10.1007/s00223-025-01356-y

Sergi, C., Benstz, J., Feist, D., Nutzenadel, W., Otto, H. F., & Hofmann, W. J. (2008). Bile duct to portal space ratio and ductal plate remnants in liver disease of infants aged less than 1 year. *Pathology, 40*(3), 260–267. https://doi.org/10.1080/00313020801911538

Shajani-Yi, Z., Ayala-Lopex, N., Black, M., & Dahir, K. M. (2020). SUN-LB64 urine phosphoethanolamine is an underutilized biomarker for hypophosphastasia. *Journal of the Endocrine Society, 4*(Suppl. 1). https://doi.org/10.1210/jendso/bvaa046.2335. SUN-LB64.

Sharma, D., Khan, I. A., Zade, A., & Sapkale, B. (2026). Alagille syndrome: A narrative review of molecular pathogenesis, clinical manifestations, diagnosis and management. *Journal of Clinical and Diagnostic Research*. https://doi.org/10.7860/JCDR/2026/82651.22320

Shea, J. L., Diamandis, E. P., Hoffman, B., Lo, Y. D., Canick, J., & Van Den Boom, D. (2013). A new era in prenatal diagnosis: The use of cell-free fetal DNA in maternal circulation for detection of chromosomal aneuploidies. *Clinical Chemistry, 59*(8), 1151–1159. https://doi.org/10.1373/clinchem.2012.201996

Tournis, S., Yavropoulou, M., Polyzos, S., & Doulgeraki, A. (2021). Hypophosphatasia. *Journal of Clinical Medicine, 10*(23), 5676. https://doi.org/10.3390/jcm10235676

Tretter, J. T., & McElhinney, D. B. (2018). Cardiac, aortic, and pulmonary vascular involvement in Alagille syndrome. In B. M. Kamath, & K. M. Loomes (Eds.), *Alagille syndrome* (pp. 77–90). Springer International Publishing. https://doi.org/10.1007/978-3-319-94571-2_6

Turnpenny, P. D., & Ellard, S. (2012). Alagille syndrome: Pathogenesis, diagnosis and management. *European Journal of Human Genetics, 20*(3), 251–257. https://doi.org/10.1038/ejhg.2011.181

Vandriel, S. M., Li, L., She, H., et al. (2023). Natural history of liver disease in a large international cohort of children with Alagille syndrome: Results from the GALA study. *Hepatology, 77*(2), 512–529. https://doi.org/10.1002/hep.32761

Vora, N. L., Langlois, S., & Chitty, L. S. (2024). Current controversy in prenatal diagnosis: The use of cfDNA to screen for monogenic conditions in low risk populations is ready for clinical use. *Prenatal Diagnosis, 44*(4), 389–397. https://doi.org/10.1002/pd.6469

Wang, E., Batey, A., Struble, C., Musci, T., Song, K., & Oliphant, A. (2013). Gestational age and maternal weight effects on fetal cell-free DNA in maternal plasma. *Prenatal Diagnosis, 33*(7), 662–666. https://doi.org/10.1002/pd.4119

Wenkert, D., McAlister, W. H., Coburn, S. P., et al. (2011). Hypophosphatasia: Nonlethal disease despite skeletal presentation in utero (17 new cases and literature review). *Journal of Bone and Mineral Research, 26*(10), 2389–2398. https://doi.org/10.1002/jbmr.454

Whyte, M. P. (2017). Hypophosphatasia: An overview for 2017. *Bone, 102*, 15–25.

Whyte, M. P., Rockman-Greenberg, C., Ozono, K., et al. (2016). Asfotase Alfa treatment improves survival for perinatal and infantile hypophosphatasia. *The Journal of Clinical Endocrinology & Metabolism, 101*(1), 334–342. https://doi.org/10.1210/jc.2015-3462

Whyte, M. P., Simmons, J. H., Moseley, S., et al. (2019). Asfotase alfa for infants and young children with hypophosphatasia: 7 year outcomes of a single-arm, open-label, phase 2 extension trial. *The Lancet Diabetes & Endocrinology, 7*(2), 93–105. https://doi.org/10.1016/S2213-8587(18)30307-3

Woolfenden, A. R., Albers, G. W., Steinberg, G. K., Hahn, J. S., Johnston, D. C. C., & Farrell, K. (1999). Moyamoya syndrome in children with alagille syndrome: Additional evidence of a vasculopathy. *Pediatrics, 103*(2), 505–508. https://doi.org/10.1542/peds.103.2.505

Zaami, S., Orrico, A., Signore, F., Cavaliere, A. F., Mazzi, M., & Marinelli, E. (2021). Ethical, legal and social issues (ELSI) associated with non-invasive prenatal testing: Reflections on the evolution of prenatal diagnosis and procreative choices. *Genes, 12*(2), 204. https://doi.org/10.3390/genes12020204

CHAPTER

9 Genetic disorders affecting growth and development: Prader–Willi syndrome and neurofibromatosis type 1

Vijayalakshmi Srinivasan[1] and Riyaz Ahmad Rather[2]

[1]*MCTM Chidambram Chettyar International School, Chennai, Tamil Nadu, India;* [2]*Department of Bioinformatics, Faculty of Engineering & Technology, Marwadi University, Rajkot, Gujarat, India*

1. Introduction

Genetic disorders constitute a major proportion of congenital and childhood-onset conditions that contribute significantly to morbidity, mortality, and long-term disability worldwide. These disorders arise from a broad spectrum of genomic alterations. Such alterations include chromosomal aneuploidies, submicroscopic copy number variations, imprinting defects, and pathogenic variants in single genes. Any of these alterations have determinantal effect on the regulation of cellular proliferation, differentiation, endocrine signaling, and neurodevelopmental pathways (Strachan & Read, 2018; Turnpenny & Ellard, 2022). Clinically, affected individuals present with growth retardation or overgrowth, hypotonia, cognitive impairment, behavioral abnormalities, dysmorphic features, and multisystem involvement. This shows the complex and pleiotropic nature of the underlying genetic mechanisms occurring in these anomalies.

Among the diverse group of genetic conditions influencing growth and development, Prader–Willi syndrome (PWS) and Neurofibromatosis type 1 (NF 1) have relatively high prevalence owing to their broad phenotypic spectra. PWS is a prototypical imprinting disorder resulting from the absence of expression of paternally inherited genes within the chromosome 15q11–q13 region, leading to hypothalamic dysfunction, neonatal hypotonia, hyperphagia, obesity, short stature, and neurodevelopmental impairment (Butler et al., 2019). In contrast, NF 1 is an autosomal dominant tumor predisposition syndrome caused by pathogenic variants in the NF 1 gene on chromosome 17q11.2, encoding neurofibromin. The latter is a negative regulator of the RAS-MAPK signaling pathway. NF 1 is characterized by café-au-lait macules, neurofibromas, skeletal abnormalities, learning difficulties, and variable effects on growth and pubertal development (Gutmann et al., 2017; Kehrer-Sawatzki et al., 2020).

Non-Invasive Prenatal Screening of Rare Fetal Genetic Diseases. https://doi.org/10.1016/B978-0-443-27660-6.00011-0

Accurate diagnosis of these disorders is important. Timely interventions can substantially improve clinical outcomes and quality of life (Deal et al., 2013; Evans et al., 2017). Definitive diagnosis is based on invasive prenatal testing procedures or postnatal molecular confirmation using karyotyping, fluorescence in situ hybridization, methylation analysis, and targeted gene sequencing. While these approaches remain the diagnostic gold standard, they are limited by procedure-related risks and restricted accessibility in certain healthcare settings. The emergence of non-invasive prenatal screening (NIPS) based on analysis of cell-free fetal DNA (cffDNA) in maternal plasma has evolved in prenatal genetics. NIPS is highly sensitive and detects fetal chromosomal abnormalities with minimal risk association (Bianchi & Chiu, 2018). NIPS has rapidly expanded to include microdeletion syndromes, imprinting disorders, and selected single-gene conditions through massively parallel sequencing (Norton et al., 2015; Zhang et al., 2019). These technological advances raise important opportunities for the early identification of PWS and NF 1 during pregnancy. This chapter focuses on the genetic basis, diagnostic strategies, and emerging role of NIPS in the context of PWS and NF 1, while critically comparing non-invasive approaches with conventional diagnostic methods.

2. Background on genetic disorders affecting growth and development

Human growth is governed by tightly regulated genetic networks that coordinate cellular proliferation, differentiation, metabolic homeostasis, endocrine signaling, and tissue patterning across prenatal and postnatal life. Perturbations in these networks lead to a wide spectrum of developmental abnormalities. Such abnormalities range from isolated growth disturbances to multisystem syndromes with neurocognitive and structural consequences. Genetic disorders affecting growth and development represent one of the most common categories of inherited disease encountered in pediatric and prenatal medicine, with an estimated prevalence of 3%–5% of live births exhibiting a major congenital anomaly attributable to genetic causes (Boyle et al., 2018; Turnpenny & Ellard, 2022). At the molecular level, these disorders arise through diverse pathogenic mechanisms. Chromosomal aneuploidies and large structural rearrangements disrupt gene dosage across multiple loci. Similarly, submicroscopic copy number variations (CNVs) may affect discrete genomic segments critical for developmental regulation. In addition, single-nucleotide variants in coding or regulatory regions can alter protein function or gene expression, while epigenetic abnormalities such as imprinting defects lead to parent-of-origin–specific gene silencing or activation (Strachan & Read, 2018). Collectively, these mechanisms underscore that growth impairment may result not only from loss-of-function mutations but also from dysregulated dosage-sensitive pathways and aberrant transcriptional control.

Endocrine pathways play a key role in somatic growth, particularly those involving growth hormone (GH), insulin-like growth factor 1 (IGF-1), thyroid hormone, and gonadotropins. Genetic disruption of these hormonal axes, or of upstream hypothalamic regulatory centers, produces characteristic patterns of short stature, abnormal body composition, and delayed or precocious puberty (Wit et al., 2019). Importantly, many syndromic growth disorders also involve primary defects in neurodevelopmental genes, reflecting the shared molecular foundations of somatic and neural maturation. Consequently, intellectual disability, behavioral disturbances, and neuropsychiatric manifestations frequently coexist with abnormal growth trajectories.

Imprinting disorders constitute an important subclass of growth-related genetic conditions. Genomic imprinting refers to the epigenetic marking of specific genes such that expression occurs from only one parental allele. Disruption of this monoallelic expression—through deletions, uniparental disomy, or imprinting center defects—results in dosage imbalance and characteristic syndromic phenotypes. Classic examples include PWS, Angelman syndrome, and Beckwith–Wiedemann syndrome, each exhibiting distinct growth patterns that reflect the functional roles of imprinted genes in energy balance or placental development (Eggermann et al., 2015; Peters, 2014).

Single-gene disorders affecting intracellular signaling pathways also represent a major category of growth and developmental syndromes. Many of these conditions involve genes that regulate mitogen-activated protein kinase (MAPK), PI3K-AKT, or mTOR pathways. The pathways are key to cell growth, survival, and differentiation. Collectively referred to as "RASopathies," these disorders, including NF 1, Noonan syndrome, and Costello syndrome, exhibit overlapping features such as short stature, craniofacial dysmorphism, skeletal anomalies, and variable cognitive impairment (Tidyman & Rauen, 2016). The convergence of phenotypes across distinct genetic lesions highlights the significance of shared molecular cascades in human development. From a clinical perspective, genetic disorders affecting growth and development demonstrate remarkable phenotypic heterogeneity, even among individuals harboring identical pathogenic variants. This variability reflects the influence of modifier genes, mosaicism, epigenetic context, and environmental factors (Cooper et al., 2013). Such complexity complicates diagnosis and underscores the limitations of phenotype-driven testing alone. Increasingly, genotype-first approaches using high-throughput sequencing and genome-wide screening have become essential for comprehensive evaluation.

Advances in genomic technologies have reshaped the conceptual framework for these disorders, shifting from single-disease models toward pathway-based and network-based interpretations. This paradigm recognizes that disturbances in shared biological systems—rather than isolated genes—drive overlapping clinical phenotypes. Within this context, PWS and NF 1 represent typical examples of how distinct molecular mechanisms (imprinting disruption versus tumor suppressor gene dysfunction) can converge on common outcomes such as impaired growth, neurodevelopmental delay, and multisystem involvement. Understanding this broader biological landscape is fundamental for appreciating the rationale behind modern diagnostic strategies, including the expanding role of noninvasive prenatal screening. Molecular genetic architecture of PWS and NF 1 are presented in Table 9.1.

Table 9.1 Molecular genetic architecture of Prader–Willi syndrome and neurofibromatosis type 1.

Feature	Prader–Willi syndrome (PWS)	Neurofibromatosis type 1 (NF1)
Chromosomal location	15q11–q13 (imprinted region)	17q11.2
Gene(s) involved	Multiple imprinted genes (SNRPN, SNORD116 cluster, MAGEL2, etc.)	NF1 gene
Gene function	Hypothalamic development, appetite regulation, neurodevelopment	Encodes neurofibromin, negative regulator of RAS signaling
Primary molecular mechanisms	Paternal deletion, maternal UPD, imprinting center defect	Missense, nonsense, frameshift, splice-site variants; large deletions
Dosage sensitivity	High (parent-of-origin dependent)	Haploinsufficiency
Mutation spectrum size	Limited recurrent regions	Thousands of private variants
Mosaicism frequency	Low	High (segmental NF1)
Genotype–phenotype correlation	Partial	Weak
Recurrence risk patterns	Depends on mechanism	50% if inherited; low if de novo
Implications for prenatal detection	CNV + epigenetic	Single-gene sequencing

3. Diagnostic approaches

Accurate diagnosis and management of genetic disorders affecting growth and development rely on cytogenetic, molecular, and genomic testing modalities (Fig. 9.1). Given the phenotypic overlap among many syndromic conditions, reliance on clinical features alone is often insufficient, particularly in the prenatal and early postnatal periods. Consequently, laboratory-based genetic testing has become essential to establish definitive diagnoses. Conventional cytogenetic analysis using G-banded karyotyping remains valuable for detecting aneuploidies, translocations, and gross deletions or duplications. However, its resolution is limited to approximately 5–10 Mb, rendering it inadequate for identifying most microdeletions, microduplications, and single-gene disorders. To overcome these limitations, fluorescence in situ hybridization (FISH) has been widely employed for targeted detection of specific genomic regions, such as the 15q11–q13 locus implicated in PWS. Although FISH offers improved resolution compared with karyotyping, it remains locus-specific and therefore unsuitable as a comprehensive screening tool (Strachan & Read, 2018).

Molecular techniques targeting epigenetic alterations have become indispensable for the diagnosis of imprinting disorders. DNA methylation analysis, using methylation-specific polymerase chain reaction (MS-PCR) or methylation-sensitive multiplex ligation-dependent probe amplification (MS-MLPA), can detect abnormal imprinting patterns at critical loci. This procedure is considered the first-line diagnostic test for PWS, identifying more than 99% of affected individuals regardless of the underlying molecular subtype (Butler et al., 2019). Subsequent assays, including SNP microarray or microsatellite analysis, are then used to distinguish among paternal deletions and maternal uniparental disomy.

Chromosomal microarray analysis (CMA) has been used to diagnose diseases that contribute to developmental delay. CMA provides genome-wide detection of copy number variations at a resolution far superior to karyotyping and has a diagnostic yield of approximately 10%–20% in unselected cohorts with neurodevelopmental disorders (Miller et al., 2010). For growth-related syndromes, CMA helps identify pathogenic microdeletions and microduplications that disrupt dosage-sensitive developmental genes.

Next-generation sequencing (NGS) technologies have further expanded diagnostic capabilities. Targeted gene panels, whole-exome sequencing (WES), and whole-genome sequencing (WGS) permit comprehensive examination of coding and noncoding regions associated with growth and developmental disorders. In the context of NF 1, sequencing of the NF 1 gene combined with deletion/duplication analysis achieves detection rates exceeding 95% in individuals fulfilling clinical criteria (Gutmann et al., 2017). More broadly, exome and genome sequencing provide diagnostic yields ranging from 25% to 40% in patients with unexplained developmental disorders, substantially surpassing traditional approaches (Clark et al., 2018).

Prenatal diagnostic pathways traditionally involve invasive sampling techniques that directly access the fetal genetic material. These procedures carry a small yet non-negligible risk and are generally offered following abnormal ultrasound findings or high-risk screening results (Wapner et al., 2012). As a result, there has been a strong impetus to develop non-invasive alternatives capable of providing comparable genetic information with minimal risk.

The discovery of cffDNA in maternal plasma has opened avenues in the development of non-invasive prenatal screening (NIPS). The technology analyzes placental-derived DNA fragments circulating in maternal blood. NIPS now encompasses detection of common aneuploidies, subchromosomal copy number changes, and selected monogenic disorders through massively parallel sequencing (Bianchi & Chiu, 2018; Zhang et al., 2019). While NIPS is currently considered a screening rather than a diagnostic test, its expanding scope has particular relevance for disorders such as PWS and NF 1, which may arise from genomic alterations theoretically detectable through cell-free DNA analysis.

4. NIPS for Prader–Willi syndrome

PWS results from the absence of expression of paternally active genes within the 15q11.2–q13 region and arises through three principal molecular mechanisms: paternal microdeletion (approximately 65%–75%), maternal uniparental disomy of chromosome 15 (25%–30%), and imprinting center defects (1%–3%) (Bittel & Butler, 2005). Because most NIPS platforms rely on low-pass whole-genome sequencing or targeted counting approaches, their primary capability is in detecting dosage abnormalities, particularly large deletions or duplications. Consequently, NIPS can potentially identify PWS when it is caused by a paternally derived 15q11.2–q13 deletion but cannot directly detect epigenetic abnormalities or UPD (Mim et al., 2024). Several reports demonstrate that whole-genome NIPS may flag abnormal chromosome 15 representation as increased risk for trisomy 15, which subsequently leads to the discovery of PWS after invasive testing. Shubina et al. (2020) described a pregnancy in which NIPS at 13 weeks indicated high risk for trisomy 15, prompting amniocentesis and SNP-based chromosomal microarray, which ultimately demonstrated maternal uniparental disomy of chromosome 15 consistent with PWS. Similarly, Ye et al. (2025) reported NIPS showing suspected trisomy 15 with a Z-score of 16.4 and fetal fraction of 11.38%, followed by amniocentesis, chromosomal microarray, trio-based whole-exome sequencing, and STR linkage analysis, which together confirmed maternal segmental UPD15 and PWS, with confined placental mosaicism for trisomy 15 identified post-termination. These observations support a mechanistic model in which NIPS does not detect PWS itself, but rather detects placental aneuploidy (most often trisomy 15), arising from meiotic nondisjunction. Subsequent "trisomy rescue" can produce a euploid fetus with maternal UPD15, resulting in PWS. Because cffDNA is largely derived from placental trophoblasts, NIPS may reflect placental mosaicism rather than fetal genotype, leading to discordance between screening and fetal molecular status.

Case-based evidence further illustrates the clinical consequences of misinterpreting such results. A neonatal case described by Japanese investigators reported an abnormal chromosome 15 Z-score on NIPS, but the absence of genetic counseling led to the continuation of pregnancy (Okuda et al., 2022). The infant was later diagnosed postnatally with PWS caused by maternal UPD derived from trisomy rescue, highlighting how failure to recognize the implications of rare autosomal trisomy signals can delay diagnosis and early intervention. The authors emphasized that NIPS platforms cannot reliably detect congenital chromosomal abnormalities such as UPD or imprinting defects and should not be considered diagnostic. Importantly, even when NIPS suggests a chromosome 15 abnormality, its diagnostic yield for PWS remains limited. While NIPS can detect copy number losses exceeding ~1 Mb, it cannot identify methylation abnormalities or distinguish among deletion, UPD, and imprinting center defects. Definitive diagnosis, therefore, requires invasive testing combined with molecular assays. Chromosomal microarray is essential for detecting deletions and regions of homozygosity suggestive of UPD. STR linkage analysis or SNP-array–based parental comparison is required to confirm parental

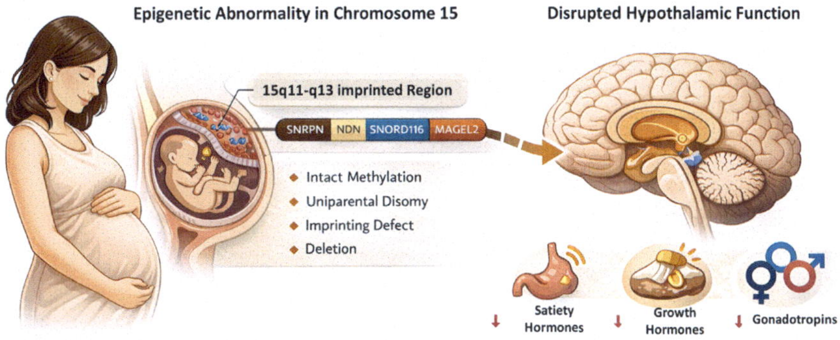

FIGURE 9.1

Epigenetic dysregulation at 15q11.2–q13 and disruption of hypothalamic function in Prader–Willi syndrome.

origin. MS-MLPA remains the most sensitive frontline test for confirming PWS, correctly identifying more than 99% of affected individuals, regardless of underlying molecular class (Brown et al., 2017). To contextualize the molecular basis underlying the limited detectability of Prader–Willi syndrome by non-invasive prenatal screening, Fig. 9.1 illustrates the epigenetic abnormalities at chromosome 15q11.2–q13 and their downstream impact on hypothalamic regulatory pathways.

From a clinical standpoint, high-risk findings for trisomy 15 or atypical chromosome 15 copy number profiles should immediately prompt genetic counseling and offer of invasive diagnostic testing. This tiered strategy is particularly critical because early diagnosis enables timely endocrine evaluation and initiation of growth hormone therapy, which has been shown to improve body composition, motor development, and cognitive outcomes in PWS (Gao et al., 2023). Thus, current evidence indicates that NIPS can occasionally enable prenatal suspicion of PWS indirectly through the detection of chromosome 15 abnormalities, particularly those related to trisomy rescue and placental mosaicism. However, NIPS lacks the capacity to directly diagnose PWS or define its molecular subtype. Its optimal role is as an entry point within a broader diagnostic pathway that mandates confirmatory invasive testing and comprehensive molecular characterization.

5. NIPS for neurofibromatosis type 1

NF1 is a single-gene condition in which definitive prenatal diagnosis requires direct or indirect detection of sequence-level variants (Fig. 9.2). Counting-based NIPS platforms designed for aneuploidy detection are therefore not suitable for NF1 screening. Instead, NIPS for NF1 relies on targeted molecular interrogation of cffDNA or intact fetal cells to determine the fetal inheritance of a known familial NF1 pathogenic variant (Gruber et al., 2018). Because confined placental mosaicism

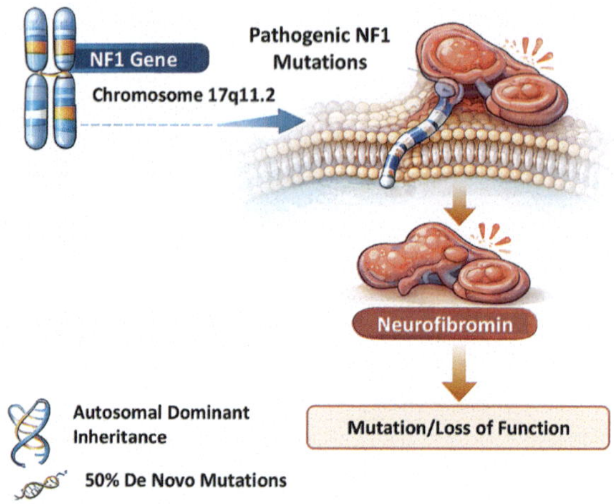

FIGURE 9.2

Genetic basis of NF1 and dysregulation of the RAS–MAPK signaling pathway.

does not affect single-gene sequence-based diagnosis in the same manner as dosage-based testing, NIPS for monogenic disorders can, in principle, provide a definitive prenatal screening without the miscarriage risk associated with invasive procedures. However, the low fetal fraction and overwhelming maternal DNA background introduce substantial technical and analytical challenges. For NF1, two major analytical strategies have emerged.

(i) **Direct variant detection**. When the disease-causing variant is paternally inherited or arises de novo in the fetus, droplet digital PCR or targeted NGS can directly detect the mutant allele in maternal plasma because it is absent from maternal DNA. This approach is highly accurate but inherently limited to situations in which the variant is not present in the mother.

(ii) **Indirect inference using relative haplotype dosage**. For maternally inherited autosomal dominant NF1 variants, direct detection is unreliable because maternal and fetal mutant alleles are indistinguishable. Relative haplotype dosage (RHDO) circumvents this by tracking parental haplotypes using dense SNP markers surrounding the NF1 locus. Over-representation or under-representation of specific haplotypes in maternal plasma reflects fetal inheritance of the mutant or wild-type chromosome. RHDO has been validated for several monogenic disorders and is increasingly regarded as the most promising strategy for maternally inherited NF1 (Pacot et al., 2024).

A large French collaborative program evaluating single-gene disorder NIPD workflows, including NF1 among target genes, demonstrated robust analytical performance using an RHDO-based framework combined with stringent quality metrics (Pacault et al., 2023). Among conclusive tests, fetal genotype was correctly assigned in 92 of 92 cases (100% specificity), while conclusive results were obtained in 92 of

98 tests (94% sensitivity). The approach was clinically viable, with a turnaround time of five to six working days, and was adaptable across mutational classes, including point mutations, deletions, duplications, and repeat expansions. These findings support the technical feasibility of NIPD for NF1 in specialized centers. Nevertheless, important biological pitfalls remain. NF1 mosaicism is common, particularly in patients with large exon deletions and certain recurrent single-nucleotide variants, and may not be detectable in peripheral blood. Mosaicism can lead to false reassurance or misinterpretation if parental mutational status is incompletely characterized prior to NIPD assay design. Moreover, germline mosaicism may result in transmission of the disease-causing haplotype even when the pathogenic variant is not detectable in maternal blood.

These complexities are well illustrated by a 5-year French prenatal diagnosis (PND) cohort comprising 205 invasive prenatal tests for NF1 (Pacot et al., 2024). Pathogenic NF1 variants were detected in 41% of fetuses, while 59% were unaffected. Importantly, two pregnancies in women with mosaic NF1 showed the absence of the maternal NF1 variant by direct testing, yet microsatellite analysis demonstrated transmission of the at-risk haplotype, highlighting the potential for false-negative results if indirect information is not incorporated. The authors emphasized that mosaicism represents a major pitfall in both prenatal and preimplantation settings and has critical implications for counseling and test interpretation.

Emerging alternatives to plasma-based cffDNA analysis include cell-based NIPT (cbNIPT), in which intact extravillous trophoblast cells are isolated from maternal blood and subjected to STR-based haplotyping and direct mutation detection (Breman et al., 2016). In a proof-of-concept study of pregnancies following preimplantation genetic testing for monogenic disorders, 89% of successfully analyzed cells were of fetal origin, and all fetal cell genotypes were concordant with invasive CVS results, supporting the potential of cbNIPT as a confirmatory or standalone approach. Although not NF1-specific, this technology is highly relevant to NF1 because it enables interrogation of large deletions, complex rearrangements, and repetitive regions that are problematic for fragmented cffDNA (Kølvraa et al., 2016). From a clinical perspective, NIPS for NF1 is currently most appropriate in families with a known pathogenic NF1 variant, where targeted assay design is possible. It is not suitable as a population screening tool and does not replace invasive testing in the absence of prior molecular characterization. Pre-test counseling must explicitly address the high de novo mutation rate, frequent mosaicism, lack of genotype–phenotype predictability, and the possibility of inconclusive results.

6. Comparison with traditional diagnostic methods

Traditional diagnostic methods for genetic disorders rely on invasive prenatal procedures and postnatal molecular testing to establish definitive diagnoses (Table 9.2). These approaches remain the gold standard for confirming disorders such as PWS and NF 1. Such methods provide comprehensive characterization of

Table 9.2 Comparative diagnostic modalities for PWS and NF1 across the prenatal–postnatal continuum.

Modality	Sample type	Detects CNVs	Detects UPD	Detects SNVs	Turnaround time	Invasiveness	PWS utility	NF1 utility
Karyotyping	CVS/Amnio	Large only	No	No	7–14 d	Invasive	Low	Low
FISH	CVS/Amnio	Targeted	No	No	1–3 d	Invasive	Moderate	Low
Methylation-PCR	CVS/Amnio/ Blood	Indirect	Yes	No	2–5 d	Invasive/ Postnatal	Gold standard	None
CMA	CVS/Amnio/ Blood	Yes	No	No	5–10 d	Invasive/ Postnatal	Subtyping	Large deletions
Targeted NGS	Blood	No	No	Yes	7–14 d	Postnatal	Rare	Gold standard
Whole exome Seq	Blood	Limited	No	Yes	3–6 wk	Postnatal	Rare	High
NIPS (counting-based)	Maternal blood	Yes	No	No	5–7 d	Non-invasive	Paternal del	None
NIPS (SNP-based)	Maternal blood	Yes	Yes	Limited	7–10 d	Non-invasive	Emerging	Targeted only

underlying molecular mechanisms with high analytical accuracy (Butler et al., 2019; Wapner et al., 2012). However, invasive sampling carries a small risk of miscarriage and is typically performed only after identification of high-risk pregnancies, limiting its applicability as a universal screening strategy. In contrast, NIPS offers a fundamentally different paradigm, emphasizing early risk assessment rather than definitive diagnosis. NIPS can be performed as early as 9–10 weeks of gestation using a maternal blood sample, providing rapid results without procedure-related fetal risk (Bianchi & Chiu, 2018). For chromosomal aneuploidies, NIPS demonstrates sensitivities and specificities exceeding 99%, but performance is more variable for subchromosomal abnormalities and monogenic disorders (Dungan et al., 2023). In the context of PWS, NIPS is primarily capable of detecting paternally derived 15q11–q13 deletions, whereas methylation-based assays identify nearly all molecular subtypes. Similarly, for NF 1, conventional sequencing of the NF 1 gene combined with deletion/duplication analysis achieves detection rates greater than 95%, whereas NIPS is currently limited to targeted detection of known familial variants or research-level genome-wide approaches.

To integrate non-invasive screening results with confirmatory molecular diagnostics, Fig. 9.3 summarizes the comparative diagnostic workflows for PWS and NF 1 across the prenatal and postnatal continuum.

CMA and sequencing-based diagnostics provide genome-wide or gene-level resolution and identify atypical or novel variants that may not be captured by targeted screening panels. NIPS, by comparison, is constrained by sequencing depth and fetal fraction, which may result in false-negative or false-positive findings, particularly for rare conditions with low population prevalence (Liehr, 2024). Additionally, confined placental mosaicism and vanishing twin can confound NIPS interpretation, whereas invasive testing analyzes fetal-derived cells more directly. From a clinical workflow perspective, traditional diagnostic testing remains indispensable for definitive diagnosis, molecular subtyping, and recurrence risk estimation. NIPS,

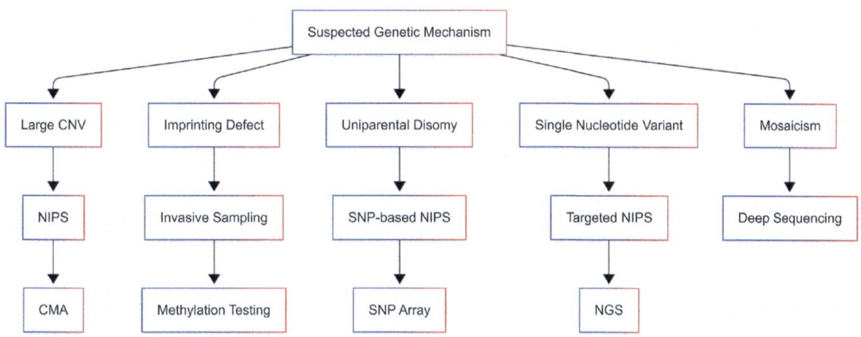

FIGURE 9.3

Integrated prenatal and postnatal diagnostic pathways for Prader–Willi syndrome and neurofibromatosis type 1.

meanwhile, occupies an expanding role in early detection and screening, complementing rather than replacing established diagnostic modalities.

7. Diagnostic yields, clinical implementation, and counseling

Across imprinting disorders, the diagnostic yield and clinical utility of non-invasive prenatal approaches have improved significantly (Table 9.3). Basically, these yields are determined by three interrelated factors: (i) biological detectability of the underlying molecular mechanism, (ii) analytical performance of the testing platform, and (iii) the availability of structured pre- and post-test counseling pathways. Evidence from recent large-scale and disorder-focused studies indicates that noninvasive technologies now achieve high analytical performance in selected contexts, but remain complementary rather than substitutive to invasive diagnostic testing. For PWS, diagnostic yield depends primarily on the molecular class. DNA methylation–based assays identify abnormal maternal-only imprinting in virtually all affected individuals, providing near-complete diagnostic sensitivity for establishing the presence of PWS, irrespective of whether the cause is deletion, maternal UPD15, or imprinting defect. However, methylation analysis alone cannot distinguish between these mechanisms, necessitating additional methods such as oligo-SNP arrays, CMA, or STR-based polymorphism testing to resolve etiology and recurrence risk. Large clinical datasets indicate that approximately 60%–70% of PWS cases are due to paternal 15q11.2–q13 deletions, 20%–30% to maternal UPD15%, and 1%–4% to imprinting defects, emphasizing the heterogeneity underlying a uniform methylation

Table 9.3 Performance characteristics of nips for rare syndromes and monogenic disorders.

Parameter	Aneuploidy	Microdeletions	Imprinting disorders	Monogenic disorders
Sensitivity	>99%	70%–95%	50%–75%	60%–90%
Specificity	>99%	98%–99%	98%–99%	98%–99%
Positive predictive value	High	Moderate	Low–moderate	Moderate
Primary limitation	Fetal fraction	CNV size	No dosage change	Variant heterogeneity
Major confounders	Placental mosaicism	Maternal CNVs	UPD	Maternal background
Confirmation required	Yes	Yes	Yes	Yes
Best clinical role	First-tier screen	Risk stratification	Selected use	Familial targeting

signature.From a non-invasive perspective, low-pass genome-wide NIPT can occasionally flag large 15q deletions or suggest trisomy 15, but it cannot directly detect UPD or imprinting defects and demonstrates an appreciable false-positive rate. A recent case of suspected trisomy 15 on NIPT ultimately proved to represent maternal UPD15 with confined placental mosaicism, confirmed only after amniocentesis, CMA, trio-based whole-exome sequencing, and STR linkage analysis. These findings underscore that, for PWS, NIPT functions primarily as a trigger for further diagnostic investigation rather than a definitive test. Consequently, the highest diagnostic yield for PWS continues to be achieved through invasive testing incorporating methylation analysis plus copy-number and polymorphism-based methods (Liang et al., 2019; Soeda et al., 2021).

In contrast, NF1 exemplifies conditions where emerging single-gene NIPT strategies can approach diagnostic-level performance in carefully selected high-risk settings. An expanded NIPS study covering 202 dominant single-gene disorders in 750 high-risk pregnancies reported a sensitivity of 100%, specificity of 99.9%, positive predictive value of 96.9%, and negative predictive value of 100%, with confirmatory whole-genome or whole-exome sequencing on amniotic fluid serving as the reference standard (Singh, 2015). Although this study was not NF1-exclusive, NF1 was included among the targeted genes, supporting the feasibility for this disorder class. Importantly, approximately 23% of confirmed positive cases involved genes not included in earlier smaller panels, demonstrating the yield advantage of expanded designs. Despite these encouraging metrics, NF1-specific prenatal diagnostic cohorts highlight biological complexities that directly affect yield.

Quality control parameters further influence diagnostic yield for all cfDNA-based assays. In a 4365-sample clinical series evaluating NIPT performance, 97.8% of samples generated successful results, while 2.2% failed quality control, predominantly because of fetal fraction below 4% (Dewantiningrum et al., 2025). Trisomy 21 remained the most frequently detected abnormality, but the study emphasized that fetal fraction and gestational age are dominant determinants of test reliability. These principles apply equally to single-gene NIPT, where low fetal fraction directly compromises the ability to detect allelic imbalance or haplotype dosage differences.

From an implementation standpoint, current evidence supports a tiered model. For PWS, invasive diagnostic testing with methylation analysis and CMA/OSA remains first-line when clinical suspicion or abnormal NIPT findings are present. For NF1 and other dominant single-gene disorders, NIPT-based approaches are best positioned as targeted tests in families with known pathogenic variants, ideally integrated within multidisciplinary prenatal diagnosis centers. Turnaround times of approximately five to six working days reported for single-gene NIPD workflows are compatible with clinical decision-making in early pregnancy.

Genetic counseling is key to maximizing clinical benefit. Pre-test counseling must clarify the distinction between screening and diagnostic testing, the disorder-specific detectability of molecular mechanisms, and the possibility of inconclusive or misleading results due to placental mosaicism (PWS) or parental mosaicism (NF1). For PWS, counseling should emphasize that recurrence risk is usually below 1% but can rise substantially in rare imprinting center defects or chromosomal rearrangements, making etiologic resolution essential. For NF1, counseling must address the 50% transmission risk in heterozygous carriers, the high de novo mutation rate, and the profound genotype–phenotype unpredictability, which limits prognostic precision even after definitive molecular diagnosis. Post-test counseling should integrate molecular results with clinical context and outline confirmatory pathways. For positive or high-risk NIPT findings, invasive confirmation remains mandatory. For negative results in high-risk families, residual risk related to mosaicism or technical limitations should be discussed. Importantly, counseling should remain nondirective and support informed parental decision-making, whether toward continuation of pregnancy, targeted neonatal management, invasive confirmation, or consideration of reproductive alternatives such as preimplantation genetic testing.

8. Case studies

Case Study 1: NIPT-Indicated Trisomy 15 Leading to Prenatal Diagnosis of PWS Caused by Maternal UPD15

A 41-year-old gravida underwent non-invasive prenatal testing at 17 weeks of gestation (Ye et al., 2025). NIPT reported an increased risk for trisomy 15. Genetic counseling was provided, and amniocentesis was performed. Chromosomal microarray analysis demonstrated long contiguous stretches of homozygosity on chromosome 15 without copy number gain or loss. Subsequent trio-based whole-exome sequencing excluded pathogenic coding variants but confirmed absence of biparental inheritance across 15q11.2–q13. Methylation-specific multiplex ligation-dependent probe amplification (MS-MLPA) showed exclusive maternal methylation patterns at the PWS critical region, confirming Prader–Willi syndrome due to maternal uniparental disomy. Placental testing after pregnancy termination demonstrated mosaic trisomy 15, supporting a trisomy-rescue mechanism. This case demonstrates that abnormal chromosome 15 signals on NIPT may represent placental mosaicism rather than true fetal trisomy, and that definitive diagnosis of PWS requires invasive testing with methylation and polymorphism-based analyses. It also illustrates the indirect but clinically valuable role of NIPT as a trigger for further diagnostic evaluation.

Case Study 2: Neonatal Diagnosis of PWS After Abnormal Chromosome 15 Z-Score on NIPT

A 34-year-old woman underwent NIPT in the first trimester, which showed negative results for trisomy 13, 18, and 21 but an atypical Z-score involving chromosome 15 (Okuda et al., 2022). No formal genetic counseling was provided, and the pregnancy continued. A male infant was delivered at 38 weeks with low birth weight and hypogonadism. Postnatal molecular testing using methylation analysis demonstrated abnormal maternal-only imprinting at 15q11.2–q13. SNP-based analysis confirmed maternal UPD15, establishing the diagnosis of PWS. This case highlights how lack of understanding of rare autosomal trisomy signals on NIPT can delay diagnosis. It emphasizes the importance of counseling and confirmatory testing whenever atypical chromosomal findings are observed, even if standard aneuploidy results are negative.

Case Study 3: Targeted Non-Invasive Prenatal Diagnosis for Familial NF1 Using Haplotype-Based Analysis

A woman with molecularly confirmed NF1 carrying a heterozygous pathogenic NF1 variant sought prenatal diagnosis (Pacault et al., 2023). Because the maternal variant could not be distinguished directly in cell-free fetal DNA, RHDO analysis was designed using informative SNPs flanking the NF1 locus. Maternal plasma sequencing demonstrated over-representation of the maternal wild-type haplotype, indicating that the fetus had not inherited the pathogenic allele. Postnatal testing of cord blood confirmed absence of the familial NF1 variant. The pregnancy was managed without invasive testing. This case illustrates the clinical feasibility of haplotype-based non-invasive prenatal diagnosis for maternally inherited NF1 variants in specialized laboratories, providing definitive fetal genotype information without procedure-related miscarriage risk.

Case Study 4: Mosaic NF1 Leading to Potential Pitfall in Prenatal Risk Assessment

A pregnant woman with clinically mild NF1 underwent blood-based molecular testing, which did not detect a pathogenic NF1 variant (Pacot et al., 2024). Based on negative maternal testing, direct variant-based NIPD was considered unnecessary. However, due to strong clinical suspicion, invasive prenatal diagnosis was performed. Microsatellite haplotype analysis demonstrated transmission of the maternal disease-associated haplotype to the fetus, and deep sequencing later identified low-level maternal mosaicism for an NF1 pathogenic variant. This case underscores that parental mosaicism can lead to false reassurance if only peripheral blood testing is used. It highlights the importance of comprehensive molecular characterization and, when appropriate, haplotype-based approaches in NF1 prenatal diagnostics.

8.1 Key lessons across cases

1. For PWS, NIST does not diagnose the disorder directly but can flag chromosome 15 abnormalities that warrant invasive confirmatory testing.
2. DNA methylation analysis remains the cornerstone of definitive PWS diagnosis.
3. For NF1, non-invasive prenatal diagnosis is most effective in families with known pathogenic variants using targeted or haplotype-based approaches.
4. Mosaicism represents a major biological limitation that must be considered in counseling and test interpretation.

Building on the diagnostic challenges illustrated by the preceding case studies, Fig. 9.4 outlines a clinically oriented decision framework linking prenatal detection to counseling, confirmatory testing, and downstream clinical management.

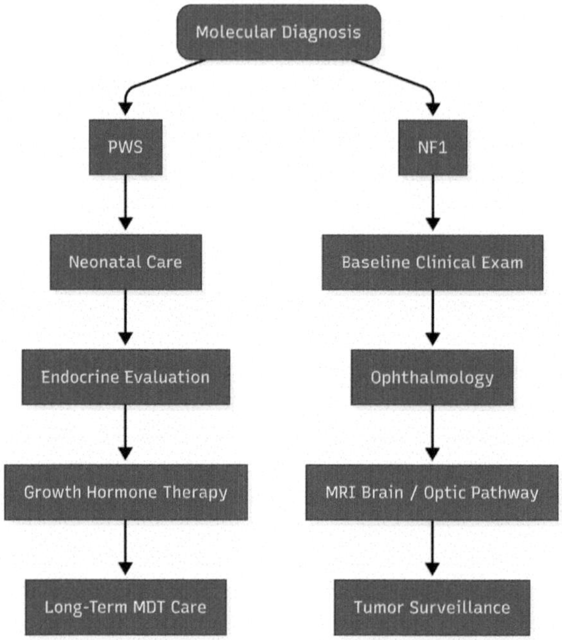

FIGURE 9.4

Clinical decision-making and counseling framework following prenatal screening of Prader–Willi syndrome and neurofibromatosis type 1.

9. Ethical and legal considerations

The integration of non-invasive prenatal screening into clinical practice for genetic disorders affecting growth and development raises complex ethical and legal questions that extend beyond technical performance. These issues include informed consent, reproductive autonomy, equity of access, data privacy, and the potential psychosocial impact of early genetic risk information. Given the expanding scope of NIPS to include rare syndromes and monogenic conditions, careful ethical oversight is essential to ensure that technological capability does not outpace responsible clinical application.

Informed consent represents a basic ethical requirement for prenatal genetic screening. Patients must receive clear, comprehensible information regarding the scope of conditions being screened, the difference between screening and diagnostic testing, the possibility of false-positive and false-negative results, and the potential need for confirmatory invasive procedures. This is particularly important for disorders such as PWS and NF 1, where NIPS does not uniformly detect all molecular subtypes. Failure to adequately convey these limitations may lead to misinterpretation of results and inappropriate clinical or reproductive decisions (American College of Obstetricians and Gynecologists, 2020; Bianchi & Chiu, 2018).

Reproductive autonomy is an essential ethical principle underpinning prenatal screening programs. The purpose of NIPS should be to provide information that supports informed decision-making, not to promote particular reproductive choices. For conditions such as NF 1, ethical concerns arise regarding the potential for stigmatization or discrimination against individuals living with the disorder. Similarly, for PWS, advances in medical management have significantly improved outcomes, complicating simplistic assumptions about quality of life.

The expansion of NIPS panels to include rare genetic disorders also raises concerns about proportionality and clinical utility. Screening for conditions with low prevalence and variable expressivity may generate more uncertainty than actionable benefit, increasing parental anxiety and healthcare costs. Ethical frameworks increasingly emphasize that conditions included in screening panels should meet criteria related to disease severity, clinical actionability, and test performance (Nuffield Council on Bioethics, 2017). Privacy and data protection represent additional ethical and legal challenges. NIPS involves large-scale genomic data generation, stored and processed by commercial laboratories. Safeguards are required to prevent unauthorized access and potential misuse of genetic information by insurers or employers. Regulatory frameworks such as data protection laws and genetic nondiscrimination legislation play a critical role in addressing these risks. Finally, issues of justice and equity must be addressed. Access to NIPS and advanced confirmatory diagnostics is uneven, both within and between countries. Without deliberate policy measures, expanding genomic technologies risks widening existing health disparities. Ethical implementation, therefore, requires parallel investment in infrastructure, training of genetic counselors, and development of context-appropriate guidelines.

10. Conclusion

Genetic disorders affecting growth and development represent a complex and heterogeneous group of conditions that impose substantial clinical and societal burden. PWS and NF 1 exemplify two distinct molecular paradigms that converge on shared clinical themes of impaired growth. Advances in molecular genetics have greatly enhanced understanding of these disorders, enabling increasingly precise diagnostic classification and more targeted clinical management. The emergence of NIPS analysis has transformed prenatal risk assessment by providing a safe, early, and widely acceptable means of detecting fetal genetic abnormalities. For PWS, NIPS is primarily effective for identifying paternally derived 15q11–q13 deletions, while emerging SNP-based strategies offer potential for detecting uniparental disomy in selected settings. For NF 1, NIPS currently holds the greatest value in targeted testing for known familial variants, with broader detection of de novo pathogenic variants remaining technically challenging. Despite its expanding scope, NIPS does not replace conventional diagnostic testing. The optimal clinical model, therefore, integrates NIPS as an initial screening and triage tool within a tiered diagnostic framework, followed by confirmatory invasive testing and comprehensive genetic counseling. Future progress in sequencing technologies, bioinformatic analysis, and fetal fraction enrichment is expected to further enhance the sensitivity and specificity of noninvasive approaches for both imprinting disorders and monogenic diseases. Equally important will be the development of evidence-based guidelines, robust counseling infrastructures, and ethical oversight to ensure responsible implementation.

References

American College of Obstetricians and Gynecologists, & Society for Maternal-Fetal Medicine. (2020). Screening for fetal chromosomal abnormalities: ACOG practice bulletin, number 226. *Obstetrics & Gynecology, 136*(4), e48–e69.

Bianchi, D. W., & Chiu, R. W. K. (2018). Sequencing of circulating cell-free DNA during pregnancy. *New England Journal of Medicine, 379*(5), 464–473.

Bittel, D. C., & Butler, M. G. (2005). Prader–Willi syndrome: Clinical genetics, cytogenetics and molecular biology. *Expert Reviews in Molecular Medicine, 7*(14), 1–20.

Boyle, C. A., Boulet, S., Schieve, L. A., Cohen, R. A., Blumberg, S. J., Yeargin-Allsopp, M., Visser, S., & Kogan, M. D. (2018). Trends in the prevalence of developmental disabilities in U.S. children, 1997–2008. *Pediatrics, 127*(6), 1034–1042.

Breman, A. M., Chow, J. C., U'Ren, L., Normand, E. A., Qdaisat, S., Zhao, L., & Beaudet, A. L. (2016). Evidence for feasibility of fetal trophoblastic cell-based noninvasive prenatal testing. *Prenatal Diagnosis, 36*(11), 1009–1019.

Brown, A., Flintoff, K., Felix, C., Gillman, C., Zamanpoor, M., & Love, D. R. (2017). Prader-Willi/Angelman syndrome: A comparison study of MS-PCR and MS-MLPA. *Medical Research Archives, 5*(9).

Butler, M. G., Miller, J. L., & Forster, J. L. (2019). Prader–Willi syndrome—Clinical genetics, diagnosis and treatment approaches. *Current Pediatric Reviews, 15*(4), 207–244.

Clark, M. M., Stark, Z., Farnaes, L., Tan, T. Y., White, S. M., Dimmock, D., & Kingsmore, S. F. (2018). Meta-analysis of the diagnostic and clinical utility of genome and exome sequencing in children with suspected genetic diseases. *npj Genomic Medicine, 3*, 16.

Cooper, D. N., Krawczak, M., Polychronakos, C., Tyler-Smith, C., & Kehrer-Sawatzki, H. (2013). Where genotype is not predictive of phenotype: Towards an understanding of the molecular basis of reduced penetrance in human inherited disease. *Human Genetics, 132*(10), 1077–1130.

Deal, C. L., Tony, M., Höybye, C., Allen, D. B., Tauber, M., Christiansen, J. S., & Savendahl, L. (2013). Growth hormone research society workshop summary: Consensus guidelines for recombinant human growth hormone therapy in Prader–Willi syndrome. *Journal of Clinical Endocrinology and Metabolism, 98*(6), E1072–E1087.

Dewantiningrum, J., Andonotopo, W., Bachnas, M. A., Pramono, M. B. A., Sanjaya, I. N. H., Darmawan, E., & Kurjak, A. (2025). Insights into noninvasive prenatal testing performance: A 4365-sample analysis of diagnostic and quality control outcomes. *SBV Journal of Basic, Clinical and Applied Health Science, 8*(3), 116–124.

Dungan, J. S., Klugman, S., Darilek, S., Malinowski, J., Akkari, Y. M., Monaghan, K. G., & ACMG Board of Directors. (2023). Noninvasive prenatal screening (NIPS) for fetal chromosome abnormalities in a general-risk population: An evidence-based clinical guideline of the American College of Medical Genetics and Genomics (ACMG). *Genetics in Medicine, 25*(2), Article 100336.

Eggermann, T., Perez de Nanclares, G., Maher, E. R., Temple, I. K., Tümer, Z., Monk, D., Mackay, D. J. G., Grønskov, K., Riccio, A., Linglart, A., & Netchine, I. (2015). Imprinting disorders: A group of congenital disorders with overlapping patterns of molecular changes affecting imprinted loci. *Clinical Epigenetics, 7*, 123.

Evans, D. G., Salvador, H., Chang, V. Y., Erez, A., Voss, S. D., Druker, H., Scott, H. S., & Tabori, U. (2017). Cancer and central nervous system tumor surveillance in pediatric neurofibromatosis 1. *Clinical Cancer Research, 23*(12), e46–e53.

Gao, Y., Zhong, M. L., Dai, Y. L., Jiang, Y. H., & Zou, C. C. (2023). Overview of genetic testing in Prader-Willi syndrome. *Expert Opinion on Orphan Drugs, 11*(1), 16–25.

Gruber, A., Pacault, M., El Khattabi, L. A., Vaucouleur, N., Orhant, L., Bienvenu, T., & Nectoux, J. (2018). Non-invasive prenatal diagnosis of paternally inherited disorders from maternal plasma: Detection of NF1 and CFTR mutations using droplet digital PCR. *Clinical Chemistry and Laboratory Medicine, 56*(5), 728–738.

Gutmann, D. H., Ferner, R. E., Listernick, R. H., Korf, B. R., Wolters, P. L., & Johnson, K. J. (2017). Neurofibromatosis type 1. *Nature Reviews Disease Primers, 3*, Article 17004.

Kølvraa, S., Singh, R., Normand, E. A., Qdaisat, S., van den Veyver, I. B., Jackson, L., & Beaudet, A. L. (2016). Genome-wide copy number analysis on DNA from fetal cells isolated from the blood of pregnant women. *Prenatal Diagnosis, 36*(12), 1127–1134.

Kehrer-Sawatzki, H., Cooper, D. N., & Kluwe, L. (2020). Mosaicism in neurofibromatosis type 1. *Human Genetics, 139*(2), 119–129.

Liang, D., Cram, D. S., Tan, H. U., Linpeng, S., Liu, Y., Sun, H., & Wu, L. (2019). Clinical utility of noninvasive prenatal screening for expanded chromosome disease syndromes. *Genetics in Medicine, 21*(9), 1998–2006.

Liehr, T. (2024). *Screening (NIPS). Non-invasive prenatal screening (NIPS) in clinical practice* (Vol 241).

Miller, D. T., Adam, M. P., Aradhya, S., Biesecker, L. G., Brothman, A. R., Carter, N. P., Church, D. M., Crolla, J. A., Eichler, E. E., Epstein, C. J., et al. (2010). Consensus statement: Chromosomal microarray is a first-tier clinical diagnostic test for individuals with developmental disabilities or congenital anomalies. *The American Journal of Human Genetics, 86*(5), 749–764.

Mim, R. A., Soorajkumar, A., Kosaji, N., Rahman, M. M., Sarker, S., Karuvantevida, N., & Uddin, M. (2024). Expanding deep phenotypic spectrum associated with atypical pathogenic structural variations overlapping 15q11–q13 imprinting region. *Brain and Behavior, 14*(4), Article e3437.

Norton, M. E., Jacobsson, B., Swamy, G. K., Laurent, L. C., Ranzini, A. C., Brar, H., Tomlinson, M. W., Pereira, L., Spitz, J. L., & Hollemon, D. (2015). Cell-free DNA analysis for noninvasive examination of trisomy. *New England Journal of Medicine, 372*(17), 1589–1597.

Nuffield Council on Bioethics. (2017). *Non-invasive prenatal testing: Ethical issues.* Nuffield Council on Bioethics.

Okuda, T., Moroto, M., & Yamamoto, T. (2022). Noninvasive prenatal testing suggesting an abnormality in chromosome 15 confirmed to be a case of Prader–Willi syndrome caused by trisomy rescue in the neonatal period. *Journal of Obstetrics and Gynaecology Research, 48*(8), 2214–2218.

Pacault, M., Verebi, C., Champion, M., Orhant, L., Perrier, A., Girodon, E., & Nectoux, J. (2023). Non-invasive prenatal diagnosis of single gene disorders with enhanced relative haplotype dosage analysis for diagnostic implementation. *PLoS One, 18*(4), Article e0280976.

Pacot, L., Vidaud, D., Ye, M., Chansavang, A., Coustier, A., Maillard, T., & Pasmant, E. (2024). Prenatal diagnosis for neurofibromatosis type 1 and the pitfalls of germline mosaics. *npj Genomic Medicine, 9*(1), 41.

Peters, J. (2014). The role of genomic imprinting in biology and disease: An expanding view. *Nature Reviews Genetics, 15*(8), 517–530.

Shubina, J., Barkov, I. Y., Stupko, O. K., Kuznetsova, M. V., Goltsov, A. Y., Kochetkova, T. O., & Sukhikh, G. T. (2020). Prenatal diagnosis of Prader-Willi syndrome due to uniparental disomy with NIPS: Case report and literature review. *Molecular Genetics and Genomic Medicine, 8*(10), Article e1448.

Singh, O. N. (2015). Clinical experience and follow-up with large-scale single-nucleotide polymorphism–based noninvasive prenatal aneuploidy testing. *American Journal of Obstetrics and Gynecology, 213*(2), 253.

Soeda, S., Saito, R., Fujii, A., Tojo, S., Tokumura, Y., & Taniura, H. (2021). Abnormal DNA methylation in pluripotent stem cells from a patient with Prader-Willi syndrome results in neuronal differentiation defects. *Stem Cell Research, 53*, Article 102351.

Strachan, T., & Read, A. (2018). *Human molecular genetics* (5th ed.). Garland Science.

Tidyman, W. E., & Rauen, K. A. (2016). Pathogenetics of the RASopathies. *Human Molecular Genetics, 25*(R2), R123–R132.

Turnpenny, P. D., & Ellard, S. (2022). *Emery's elements of medical genetics and genomics* (16th ed.). Elsevier.

Wapner, R. J., Martin, C. L., Levy, B., Ballif, B. C., Eng, C. M., Zachary, J. M., Savage, M., Platt, L. D., Saltzman, D., Grobman, W. A., et al. (2012). Chromosomal microarray versus karyotyping for prenatal diagnosis. *New England Journal of Medicine, 367*(23), 2175–2184.

Wit, J. M., Kiess, W., & Mullis, P. E. (2019). Genetic evaluation of short stature. *Best Practice and Research Clinical Endocrinology and Metabolism, 33*(3), Article 101277.

Ye, Y., Fu, Y., Zhang, Z., Ning, H., Tao, F., Wang, X., & Hao, X. (2025). Prenatal diagnosis of Prader–Willi syndrome via maternal UPD15 with placental mosaicism: Incidental discovery of fetal DMD carrier status. *Frontiers in Genetics, 16*, Article 1675663.

Zhang, J., Li, J., Saucier, J. B., Feng, Y., Jiang, Y., Sinson, J., McCombs, A. K., Schmitt, E. S., Peacock, S., Chen, S., Tian, X., Song, F., & Li, X. (2019). Non-invasive prenatal sequencing for multiple Mendelian monogenic disorders using circulating cell-free fetal DNA. *Nature Medicine, 25*(3), 439–447.

Further reading

Cassidy, S. B., Schwartz, S., Miller, J. L., & Driscoll, D. J. (2012). Prader–Willi syndrome. *Genetics in Medicine, 14*(1), 10–26.

Chitty, L. S., Mason, S., Barrett, A. N., McKay, F., Lench, N., Daley, R., Jenkins, L., Madgett, T., Firth, H. V., & Lo, Y. M. D. (2018). Non-invasive prenatal diagnosis of achondroplasia and thanatophoric dysplasia: Next-generation sequencing allows for a safer, more accurate approach. *Prenatal Diagnosis, 38*(1), 30–37.

Dar, P., Curnow, K. J., Gross, S. J., Hall, M. P., Stosic, M., Demko, Z., Zimmermann, B., Hill, M., Sigurjonsson, S., Ryan, A., & Rabinowitz, M. (2014). Clinical experience and follow-up with large scale single-nucleotide polymorphism-based non-invasive prenatal aneuploidy testing. *American Journal of Obstetrics and Gynecology, 211*(5), 527. e1–527.e17.

Fan, H. C., Gu, W., Wang, J., Blumenfeld, Y. J., El-Sayed, Y. Y., & Quake, S. R. (2012). Non-invasive prenatal measurement of the fetal genome. *Nature, 487*(7407), 320–324.

Kitzman, J. O., Snyder, M. W., Ventura, M., Lewis, A. P., Qiu, R., Simmons, L. E., Gammill, H. S., Rubens, C. E., Santillan, D. A., Murray, J. C., & Shendure, J. (2012). Noninvasive whole-genome sequencing of a human fetus. *Science, 336*(6085), 149–152.

CHAPTER

10 Ethical considerations in non-invasive prenatal screening

Anam Mursaleen and Javaid Ahmad Sheikh
Department of Biotechnology, Jamia Hamdard, New Delhi, India

1. Introduction

Non-invasive prenatal testing (NIPT), also called non-invasive prenatal screening (NIPS), refers to the analytical examination of cell-free DNA fragments in the blood of pregnant women to determine the probability that the fetus has certain chromosomal conditions. The scientific basis for NIPS was established by Lo and colleagues, who identified that fetal DNA, mostly placental, can be found in the plasma and serum of pregnant women (Lo et al., 1997). This observation allowed the development of non-invasive fetal genetic evaluation via simple maternal blood sample collection. The evolving method has since transformed pregnancy screening paradigms. In comparison with amniocentesis and chorionic villus sampling, NIPS has no risk and allows early detection, as early as 9 or 10 weeks into a pregnancy. The analytic sensitivity and specificity of NIPS for the common aneuploidies, including Down syndrome, or trisomy 21; Edwards syndrome, or trisomy 18; and Patau syndrome, or trisomy 13, are exceptionally high in high-risk populations (Bowman-Smart et al., 2020; Zaami et al., 2021).

However, NIPS diverges from the prenatal screening that precedes it in at least three ethical dimensions. First, it is low-burden and potentially routine, increasing the risk that testing proceeds with diminished deliberation. The Royal College of Obstetricians and Gynecologists (RCOG) has already warned about the risk of not raising the issue of miscarriage risk during prenatal testing and the need for particularly strong communication about the results of the test, which is seen as "a simple, safe blood test". Second, NIPS is a probabilistic test, and even with good analytic performance, the clinical significance of a positive result is dependent upon the baseline prevalence of the condition and the positive predictive value, with the possibility of false positive and false negative results. The American College of Obstetricians and Gynecologists (ACOG) clearly emphasizes the distinction between screening and diagnostic testing, and counseling that incorporates predictive value information is advocated (ACOG, 2020). Third, NIPS has the potential for information overload, given the rapid expansion of the scope of NIPS, including sex chromosome aneuploidies, microdeletions, and genome-wide copy number variation, and the uncertain clinical utility of such

Non-Invasive Prenatal Screening of Rare Fetal Genetic Diseases. https://doi.org/10.1016/B978-0-443-27660-6.00001-8

information, which has been addressed by the Society for Maternal-Fetal Medicine (SMFM) by providing updated recommendations that clearly outline the conditions under which NIPS for specific conditions is indicated or not indicated (SMFM, 2025). Furthermore, the ease and safety of NIPS have driven up uptake around the world. In a handful of rich countries, it is now used in 25%–50% of pregnancies, and parts of Europe even incorporate their NIPS uptake into national screening programmes. In India, no formal nationwide screening program exists as of yet. However, evidence suggests that pregnant women are interested-one survey suggested 83% of Indian women would screen prenatally if their physician recommended it, and 85% would support a national screening program (Arumugam et al., 2024). Nevertheless, NIPS in India is primarily conducted through private laboratories, and frequently funded out-of-pocket (Jayashankar et al., 2023), leading to the uptake being among urban and socio-economically advantaged groups, raising questions of distributive justice. At the same time, NIPT's ability to produce genetic information early in the pregnancy raises complex ethical, legal, and social issues (ELSI) (Zaami et al., 2021). In principle, NIPTs provide expectant parents with information to guide reproductive decisions, thereby enabling autonomy and reducing risks. For example, the test consistently discloses the sex of the fetus and has heightened anxiety regarding sex-selective abortion, especially in countries such as India that have a well-documented gender preference (Bowman-Smart et al., 2020). Moreover, the prevalence of screening for conditions like Down syndrome generates concern regarding the inherent value of people with disabilities, which may communicate that such lives are less "valuable," reinforcing stigma and discrimination (Kater-Kuipers et al., 2020). Fig. 10.1 shows the ethical framework used in NIPS.

From an ethical perspective, NIPS appears to exist in a complex normative space. NIPS is located in a normative space that is complex and challenging. It offers significant benefits, such as early and safe access to genetic risk information, while it demands that healthcare systems protect informed consent, prevent diagnostic drift, ensure equitable access, prevent misuse, and maintain the dignity of individuals with disabilities. The normative issue with NIPS revolves around balancing reproductive autonomy with the risk of harm, coercion, inequity, and discrimination. This chapter begins with an overview of the main ethical principles of utilitarianism, deontology, and virtue ethics, and then examines the implications of prenatal screening in each of these ethical principles. This chapter also examines and critically discusses some of the main ethical issues in prenatal screening, including informed consent, probabilistic reasoning, privacy, equity of access, disabilities, and cultural issues such as sex selection in India. The chapter also outlines a balancing framework integrating the international professional consensus with the Indian legal framework of the Medical Termination of Pregnancy Act and the PCPNDT Act to outline the ethical implications of non-invasive prenatal screening in various settings.

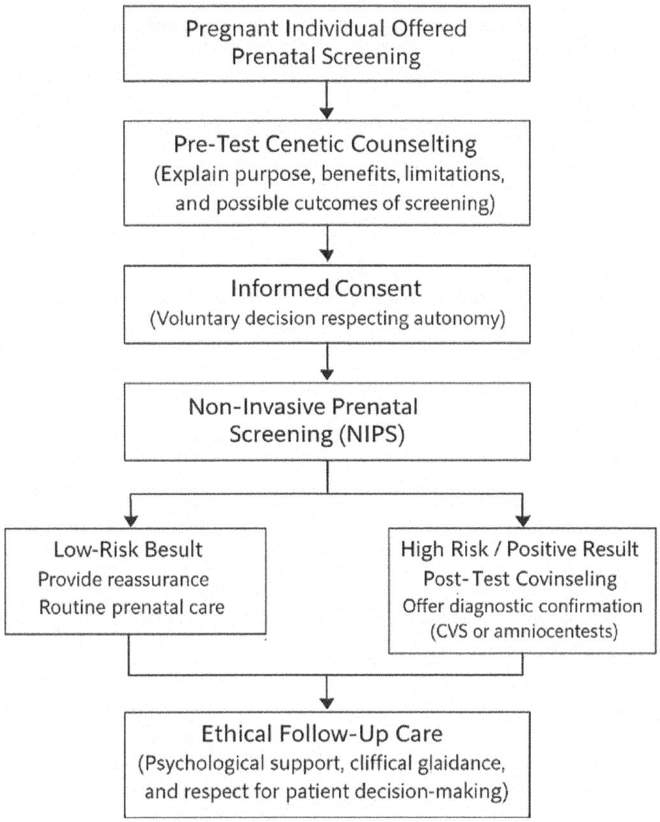

FIGURE 10.1 Ethical implementation pathway for NIPS

This schematic flowchart illustrates the ethically responsible clinical pathway for the use of non-invasive prenatal screening. The process begins with pre-test counseling and informed consent, followed by screening and interpretation of results. Low-risk results typically lead to routine prenatal care, whereas high-risk results require post-test counseling and confirmatory diagnostic testing.

2. Ethical frameworks in prenatal screening

Ethical evaluation of NIPS can be best anchored to the ethical principles of respect for autonomy, informed consent, beneficence, non-maleficence, justice, and privacy as depicted in Fig. 10.2. These principles have practical implications for the workflow and the way the procedure will be delivered, the way it will be extended, the way it will be followed up, and the way the data will be used. This approach also finds support from the profession, which considers counseling and follow-up as essential components of ethical NIPS and not as supplementary components (ACOG, 2020).

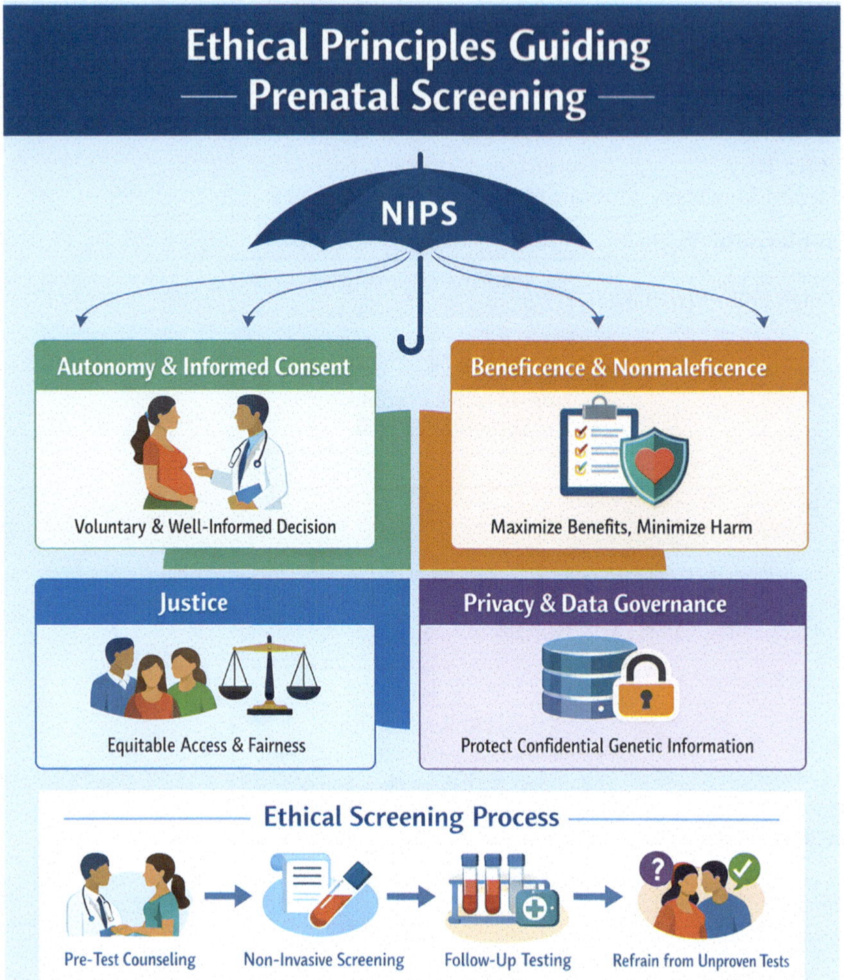

FIGURE 10.2 Ethical principles guiding NIPS. The diagram illustrates the core bioethical principles that underpin the ethical implementation of NIPS

The framework highlights four key principles: autonomy and informed consent, ensuring voluntary and well-informed reproductive decision-making; beneficence and nonmaleficence, emphasizing maximization of clinical benefits while minimizing potential harms; justice, which promotes equitable access to screening services and fair distribution of healthcare resources; and privacy and data governance, focusing on the protection of sensitive genetic information and responsible management of genomic data.

2.1 Core bioethical principles

2.1.1 Autonomy and informed consent

Autonomy demands that patients have the freedom to consent or not consent to testing without any form of coercion and must also have sufficient understanding. ACOG affirms, "All patients should be offered both screening and diagnostic tests and have the right to accept or refuse these tests after counseling." Ethical consent in NIPS must additionally incorporate understanding of probabilistic meaning (PPV, residual risk) and meaning of outcomes (high chance, low chance, no-call) as these may shape immediate reproductive decisions. ACOG notes, for example, that PPV is dependent on prevalence and that true positive risk varies across populations (ACOG, 2020).

2.1.2 Beneficence and nonmaleficence

The benefit of the NIPS is substantial in the context of unnecessary invasive procedures being avoided, and risk-stratification being provided at an earlier time based on greater accuracy. However, harm is done when false positives, false negatives, indeterminate, and "no-call" results are mismanaged, particularly if screening results lead to irreversible decisions before diagnostic testing is performed. Both ACOG and the Food and Drug Administration (FDA) emphasize the need for diagnostic testing following screen-positive patterns of screening results. ACOG outlines a framework for managing "no-call" results where there is increased risk of aneuploidy (ACOG, 2020).

2.1.3 Justice

Justice demands equitable access not only to screening services but also to counseling, confirmation, and condition-specific supports. Empirical gradients of uptake according to socioeconomic status suggest that "availability in principle" is not equivalent to "equitable access in practice" (van der Meij et al., 2021).

2.1.4 Privacy and data governance

The WHO recommendations for 2024 have defined the governance of genomic data as a lifecycle commitment, with a focus on the importance of stewardship, accountability, equity, and the prevention of misuse. For prenatal genomic screening, which involves the future individual and their family, the WHO has highlighted the importance of the child's "right to an open future" and the need for governance that supports individuals, families, and communities (ACOG, 2020).

2.2 Ethical theories applied to prenatal screening

2.2.1 Utilitarianism in prenatal screening

Utilitarianism is a consequentialist ethical theory that focuses on the outcomes of actions to achieve the greatest good. A utilitarian approach to prenatal screening would take into account the maximum benefit or harm to the population. From this perspective, cfDNA screening for common autosomal trisomies can be considered justified with the prospect of reducing the harm of the procedure, increasing the detection rate, and optimizing the use of diagnostic tests. A meta-analysis conducted by Gil et al. (2017) showed that cfDNA screening for singleton pregnancies can detect more than 99% of trisomies 21, 98% of trisomies 18%, and 99% of trisomies 13, with a false-positive rate of 0.13% (Gil et al., 2017). Furthermore, the decrease in the occurrence of serious genetic disorders could be considered to be increasing the good in the world by reducing pain and medical expenses. Therefore, prenatal screening and selective abortion could be considered to be the sensible course of action because of the possibility of creating healthier offspring and reducing the burden on society. In this regard, philosopher Peter Singer, in support of the utilitarian perspective, has argued that "the life of the mother takes precedence over the life of the fetus" and that parents have moral reason to choose the child that has the greatest prospects in life. Utilitarians can use the principle of procreative beneficence as enunciated by Savulescu, which argues that couples should choose the life with the best prospects (Maftei & Dănilă, 2022). However, the utilitarian argument can be used to justify unethical decisions if one focuses solely on outcomes rather than considering informed decisions and fairness in distribution. "Panel expansion" in utilitarianism may be detrimental in cases where increasing uncertainty and lower PPV lead to more false alarms and anxiety without improving patient outcomes. This rationale is exemplified in SMFM's "non-recommendation" for routine microdeletions in healthy populations and ACOG's warning against unvalidated genome-wide "ancillary" screening without established accuracy and false positive rates (SMFM, 2025). Also, the 2021 ethical review by Genes on prenatal screening recognizes the potential for stigmatization and discrimination of disabled people in the event of the technology being utilized without proper ethical considerations (Zaami et al., 2021). Therefore, though the utilitarian ethical principle justifies the need for NIPS in preventing certain genetic diseases, it has to consider the moral incurred in perpetuating stigma and devaluing the lives of people living with disabilities.

2.2.2 Deontology and duty-based ethics

Deontological ethics, particularly as embodied in Kantian ethics, emphasizes duty, right, and moral principles in preference to consequences. In the realm of NIPS, deontological ethics is concerned with the moral duty of healthcare practitioners in terms of their duty to inform, respect autonomy, avoid causing foreseeable harm, maintain confidentiality, and provide non-directive counseling (Barrow & Khandhar, 2026; Mandal et al., 2016). One of the primary duties of ethical NIPS is to prevent diagnostic drift, which refers to the misunderstanding of the screening test as a diagnostic test. The United States Food and Drug Administration clearly indicates that NIPS may cause false results, which should not be used for making irreversible decisions without confirmatory diagnostic testing. On the other hand, the ACOG recommended extensive counseling on screening and diagnostic options, confirmatory testing after positive screening, and handling "no call" results that indicate a high probability of aneuploidy (ACOG, 2020). Deontological ethics also emphasizes the duty to establish real informed consent. There is a duty on the part of healthcare practitioners to provide information on the potential of non-invasive prenatal screening tests, including their probabilistic nature and the possibility of false positives and false negatives, and the need to carry out diagnostic tests. There is a need to empower patients to undergo or forgo tests without any form of coercion, and this is a reflection of respect for autonomy. In this theoretical framework, the pregnant woman is an autonomous moral agent whose decisions regarding reproduction are worthy of respect, though some versions of deontology also emphasize a duty to fetal life depending on how it is construed (Minear et al., 2015). Duty-based reasoning also delimits prenatal screening. Even when patients desire a wider range of genomic testing, doctors have a duty not to inappropriately offer poorly validated or low-utility tests (Dickens, 2014). In recognition of their duty, the ACOG points to the low positive predictive value and increased false-positive rates for testing for rare microdeletions. It also states that genome-wide cell-free DNA screening lacks clinical validation for routine clinical use (Petersen et al., 2017). A deontological approach would suggest that screening should be provided without any underlying motivations, as the purpose of screening should be informative, not disallowing the birth of persons with disabilities. Therefore, NIPT should respect reproductive autonomy by providing non-directive counseling, while respecting the choices of parents and maintaining (Pertile, 2018).

2.2.3 Virtue ethics and the role of character

Virtue ethics focuses on decision-makers and their character and moral qualities, such as virtues required for coping with uncertainty and providing advice during prenatal screening. It emphasizes virtues such as being honest about risk, humble about predictive limits, empathetic toward different parental values, and having practical wisdom (phronesis) while providing advice. Clinically, virtues such as compassion, fairness, and prudence guide decision-makers toward providing sensitive advice to patients (Kotzee et al., 2017). Specific professional guidance emphasizes the relevance of these virtues in the NIPS context. The RCOG stresses that good communication is important because, if NIPT is framed as a "simple, safe blood test," it can inhibit critical engagement with its implications and undermine informed decision-making. Furthermore, the SMFM notes that the commercialization and increasing complexity of NIPS outpace clinicians' understanding and use, underlining the importance of a system supportive of sound professional judgment (ACOG, 2020; Bowman-Smart et al., 2020). Virtue ethics emphasizes the role of systemic factors that impede good practice, such as rushed counseling, inadequate training, limited referral systems, and unclear lab reports. This situation may lead to moral triage, causing distress and potential for error. Parents must also consider prudence and courage, where the implications and consequences of testing must be assessed to promote the child's well-being in the future, according to Guido Tonkens, where "virtuous parents are those who are committed to providing for a child's future flourishing" (Tonkens, 2015). In summary, virtue ethics highlights that ethical prenatal screening is not only about rules and policies but also about the character and judgment of practitioners and parents. Such a prenatal screening model based on virtue ethics encourages compassionate counseling, wise communication of uncertainty, and decision-making that respects practitioners' integrity and parents' caring.

3. Key ethical challenges

Building on the frameworks outlined above, a range of practical ethical issues arises in relation to NIPT. These include issues of consent, possible harm (physical, psychological, and social), issues of justice and equity, cultural issues (which include issues of sex selection), and issues of privacy in relation to genetic information.

3.1 Informed consent and counseling

The ease and simplicity of the test, requiring merely a blood test, and its incorporation into routine prenatal care may lead to trivialization, with the patient possibly failing to fully appreciate that it is a genetic test with significant implications. Empirical evidence shows that a majority of pregnant women undergo prenatal tests without an accurate understanding of their benefits and risks. Ethicists have also cautioned that there is a risk of the "screening trap", wherein the consent for one test may lead to pressure to perform other tests without adequate consideration of the implications of the test results. An Italian committee has cautioned that the "perception of the test as safe may lead to the decision to undergo the test being taken lightly, almost as "just another blood test." To address the ethical concerns of the trivialization of the test, professional guidelines recommend pre-test counseling as well as post-test counseling for individuals undergoing NIPT. It has been found that informed decision-making, which is well-informed and in line with the patient's values, is associated with higher satisfaction levels and decisional conflicts (van den Berg et al., 2005). In India, a lack of qualified genetic counselors is a challenge. Most obstetricians lack enough time to counsel pregnant women. Surveys have shown that pregnant women desire genetic counseling. In a survey done in India, 69% of pregnant women wanted to discuss prenatal tests with a genetic counselor. Another survey indicated that 84% of pregnant women believed that awareness of prenatal tests should be mandatory in early pregnancy (Arumugam et al., 2024).

3.2 Psychosocial impact and decisional burden

The associated psychosocial profile for NIPS is not entirely reassuring. A scoping review done by Labonté et al. suggests that only a small number of studies have evaluated the psychological consequences of NIPS, and these are generally in the short-term. However, current literature suggests that short-term anxiety is reduced in women receiving a negative result and decisional regret is low (Labonté et al., 2019). Ethically, these findings support two constraints for clinical practice: do not promise a psychological advantage, and support individuals receiving a high probability or no-call result (Johnston et al., 2023; Vanstone et al., 2018).

3.3 Privacy and genetic data

Sensitive genetic information is revealed in the course of NIPT tests. Moreover, apart from the usual conditions that are screened for, such as chromosomal abnormalities, modern technology allows for the identification of additional genetic factors or predispositions to adult-onset diseases. The issue that is raised in this respect is related to privacy. If a prenatal test unexpectedly reveals a potential health issue, such as a predisposition to Huntington's disease or a mother's potential for cancer, there is a query regarding who should be informed. The debate on the "right to an open future" is related to the issue of whether or not the parents have a right to know, or be informed, regarding a potential adult-onset disease that the child is predisposed to. The issue that is raised is related to confidentiality, and it is legally and ethically imperative that confidentiality is ensured in the case of NIPT test results, just like any other medical test result, so that there is no disclosure to third parties, such as employers or insurance companies. The issue of genetic privacy in India is still in its infancy, and there is a potential for a child being discriminated against if he or she is found to carry a potential health issue (WHO, 2024).

3.4 Non-maleficence

Among the four principles of bioethics, non-maleficence or "do no harm" is one of the most important. While NIPT does not cause physical harm to the fetus, unlike invasive tests, it can cause other types of harm. Some of these include emotional harm, such as causing undue stress and anxiety. Other potential harms include social harm, such as when NIPT results are used for sex selection or to exclude fetuses with disabilities, leading to a negative impact on society. In India, prenatal sex determination is a highly controversial topic. The PCPNDT Act of 1994 has made it illegal to reveal the sex of a child. Therefore, revealing such information is a crime. This has led to a situation wherein such information is not disclosed by ultrasound centers. However, NIPT can reveal fetal sex as early as 7–9 weeks into a pregnancy, compared to ultrasound, which can detect it after 16 weeks (Li et al., 2024). The fear is that such information can be used for sex-selective abortions, leading to an already skewed sex ratio in Indian society. A recent review has noted that the daunting prospect of NIPT results being used for sex selection in countries such as India and China has been a concern. This is a type of social harm because it goes against gender justice and can put undue pressure on a woman to abort a female child. This is a critical situation faced by healthcare professionals, wherein they have to perform a balancing act between providing information and engaging in illegal practices. In India, a healthcare professional can withhold information regarding fetal sex and instead claim that it has been carried out to detect anomalies (Svirsky et al., 2023).

3.5 Disability rights and stigma

Disability advocates argue that the practice of prenatal screening could inadvertently promote negative messages about disability. This is because conditions such as Down syndrome are common targets for prenatal screening, and the practice of conducting wide-scale screenings could inadvertently promote the notion that the lives of individuals with disabilities are less desirable. Empirical studies have confirmed that prenatal screenings reduce the birth rate of individuals with disabilities (de Jong & de Wert, 2015). Ethical analyses indicate that prenatal screening should promote informed reproductive choice rather than aiming to reduce the birth of disabled people, as pointed out by Annemarie Kater-Kuipers (Kater-Kuipers et al., 2018). The Nuffield Council on Bioethics argues about the wider ethical implications of the increased use of NIPS, affecting how society perceives pregnancy, disability, and human difference, and therefore distorts stigma and social meaning out of social and cultural norms. These issues are of particular concern in India, where support systems for persons with disabilities tend to be limited. Ethical practice also entails the provision of information that is balanced and disability-inclusive, and the demonstration of commitment to disability support systems, ensuring that reproductive choices are not based on stigma and lack of support systems (Allyse et al., 2013).

3.6 Regulatory and legal constraints

The regulation influences ethical considerations by governing marketing claims, reporting practices, and the need for clinical follow-up based on screening test results. For example, the regulation emphasizes that NIPS is a screening test rather than a diagnostic test, and that there is a risk of false results or misinterpretation if the test is acted upon without confirmation by a diagnostic test. Biological factors, such as confined placental mosaicism, may also result in discordant or false-positive results, highlighting the need for careful counseling and follow-up through diagnostic procedures (Mardy & Wapner, 2016; Wu et al., 2021). In India, the ethical landscape surrounding NIPS is shaped by complementary regulations regarding diagnostic tests and abortion. In addition to the law against prenatal sex selection, the Medical Termination of Pregnancy Act dictates the boundaries of legal abortion. With the 2021 amendment, abortion is permitted up to a gestation of 24 weeks in specific circumstances, where previously, without the amendment, termination of pregnancy was legal up to 20 weeks. NIPT is typically performed around 10–12 weeks of gestation and can provide information on genetic risk early enough to allow for confirmatory testing and legally permissible decisions regarding pregnancy management. Regulatory oversight is a key factor in global ethical practices. The FDA has issued safety alerts regarding NIPS, including false results, misinterpretation, and decision-making without confirmatory testing. In the UK, it is through public health programs, including the NHS Fetal Anomaly Screening Program (FASP), using contingent thresholds, along with access to confirmatory tests such as Chorionic Villus Sampling/Amniocentesis, after high-probability results. These frameworks demonstrate how legal and program design impact ethical prenatal screening practices (Chedid & Phillips, 2019; Dondorp & van Lith, 2015; Potter et al., 2008).

4. Balancing framework

A practical balancing framework for the implementation of NIPS views screening as an important value-sensitive aspect of healthcare. It brings together the ethical principles with the guideline-based operational needs, resulting in a balanced approach for the conduct of NIPS screening.

4.1 Integrating autonomy with beneficence-based care

Autonomy is relevant when the consent is freely given and well-informed. The principles of beneficence and non-maleficence are met when the screening reduces harm, not when it causes undue anxiety or excessive investigation that is unwarranted. The ACOG and SMFM provide a detailed balancing act that includes the following steps: (1) offering choices with appropriate counseling; (2) offering screening, not a diagnostic test; (3) offering clear and structured follow-up for positive or no response results; and (4) withholding routine screening when its efficacy and usefulness are in question (Gregg et al., 2016; Skotko et al., 2019; SMFM, 2015).

4.2 Balancing justice with pathway design

Justice is achieved when the entire screening experience is considered just from the very beginning to the end, including the right counseling and support. Research indicates that the socioeconomic status of an individual can influence the decision to undergo NIPT, disrupting equal access and requiring the monitoring and addressing of inequalities (van der Meij et al., 2021). In national screening programs, the governance structure is supposed to ensure the implementation of the principle of equity. The NHS FASP takes a contingent approach to NIPT, whereby NIPT is performed following a high-risk result from first-line screening, as well as ensuring access to diagnostic procedures such as chorionic villus sampling or amniocentesis, and the necessary counseling following a high-risk result from first-line screening (Allyse et al., 2015).

4.3 Balancing privacy with clinical and research value

The balancing act between privacy concerns and the clinical and research value of genomic screening programs is an ongoing issue. The call for global governance is for transparent, fair, and accountable genomic data management across the world, with particular attention given to children's interests and their "open future" concerns. The policy recommendations issued by the World Health Organization on human genome data governance, among other general ethical principles, encourage responsible oversight, fair access, and robust protection of individual rights in relation to genomic data management for clinical or research purposes. Reality also supports this, as patients do not have an adequate understanding of their genomic data sharing for clinical or research purposes. This is an important ethical consideration that highlights that informed consent should not simply be an exercise in contractual requirements but should also involve discussions around data sharing (McGuire et al., 2011; Middleton et al., 2020; Sanderson et al., 2017).

5. Conclusion

NIPS raises complex ethical, legal, and social concerns that extend beyond its clinical utility. Informed and voluntary decision-making is the most encountered ethical interest. As NIPS is increasingly integrated into routine prenatal care, there is a risk that it may be perceived as a standard test rather than a sensitive genetic screening tool. This necessitates robust pre- and post-test counseling to ensure that reproductive choices reflect the individual's values rather than external pressures from clinicians or societal expectations. Ethical evaluation of NIPS can be understood through multiple frameworks. Utilitarian perspectives emphasize population-level benefits, such as reduced reliance on invasive procedures, but may overlook individual autonomy and equity. In contrast, deontological ethics prioritize duties such as respecting autonomy, ensuring accurate information, and avoiding harm, particularly by distinguishing screening from definitive diagnosis. Virtue ethics further highlights the importance of empathy, honesty, and professional integrity in communicating complex genetic information. Practical ethical challenges are the psychosocial impact of results, where uncertain or adverse findings may increase anxiety and decisional burden, necessitating supportive counseling. Additionally, genetic privacy and data governance have emerged as significant concerns, particularly with expanding genomic information and potential misuse. From a societal perspective, NIPS raises questions regarding disability rights and social justice. There is concern that widespread screening may reinforce stigma toward disabilities if not framed within informed choice and respect for diverse lived experiences. In the Indian context, ethical concerns are further shaped by regulatory frameworks such as the PCPNDT Act, which aims to prevent sex-selective practices, and the MTP Act governing pregnancy termination. Issues of equitable access remain critical, as high costs may limit availability to certain populations. Ethical implementation of NIPS, therefore, requires balanced governance, strong regulation, and commitment to justice, autonomy, and compassionate reproductive care.

References

Allyse, M., Minear, M. A., Berson, E., et al. (2015). Non-invasive prenatal testing: A review of international implementation and challenges. *International Journal of Women's Health, 7*, 113–126. https://doi.org/10.2147/IJWH.S67124

Allyse, M. A., Sayres, L. C., Havard, M., et al. (2013). Best ethical practices for clinicians and laboratories in the provision of noninvasive prenatal testing. *Prenatal Diagnosis, 33*(7), 656–661. https://doi.org/10.1002/pd.4144

American College of Obstetricians and Gynecologists Committee on Practice Bulletins Obstetrics, Committee on Genetics, & Society for Maternal-Fetal Medicine. (2020). Screening for fetal chromosomal abnormalities: ACOG practice bulletin, number 226. *Obstetrics & Gynecology, 136*(4), e48–e69. https://doi.org/10.1097/AOG.0000000000004084

Arumugam, S., Kalluri, S. S., Sharmila, V., et al. (2024). Acceptability of prenatal screening tests among expectant mothers in India: Insights and implications for public health. *Cureus, 16*(5), Article e61246. https://doi.org/10.7759/cureus.61246

Barrow, J. M., & Khandhar, P. B. (2026). Deontology. In *StatPearls*. StatPearls Publishing. http://www.ncbi.nlm.nih.gov/books/NBK459296/.

Bowman-Smart, H., Savulescu, J., Gyngell, C., Mand, C., & Delatycki, M. B. (2020). Sex selection and non-invasive prenatal testing: A review of current practices, evidence, and ethical issues. *Prenatal Diagnosis, 40*(4), 398–407. https://doi.org/10.1002/pd.5555

Chedid, R. A., & Phillips, K. P. (2019). Best practices for the design, implementation and evaluation of prenatal health programs. *Maternal and Child Health Journal, 23*(1), 109–119.

de Jong, A., & de Wert, G. M. W. R. (2015). Prenatal screening: An ethical agenda for the near future. *Bioethics, 29*(1), 46–55. https://doi.org/10.1111/bioe.12122

Dickens, B. M. (2014). Ethical and legal aspects of noninvasive prenatal genetic diagnosis. *International Journal of Gynecology & Obstetrics, 124*(2), 181–184. https://doi.org/10.1016/j.ijgo.2013.11.001

Dondorp, W., & van Lith, J. (2015). Dynamics of prenatal screening: New developments challenging the ethical framework. *Bioethics, 29*(1), ii–v.

Gil, M. M., Accurti, V., Santacruz, B., Plana, M. N., & Nicolaides, K. H. (2017). Analysis of cell-free DNA in maternal blood in screening for aneuploidies: Updated meta-analysis. *Ultrasound in Obstetrics and Gynecology, 50*(3), 302–314. https://doi.org/10.1002/uog.17484

Gregg, A. R., Skotko, B. G., Benkendorf, J. L., et al. (2016). Noninvasive prenatal screening for fetal aneuploidy, 2016 update: A position statement of the American College of Medical Genetics and Genomics. *Genetics in Medicine, 18*(10), 1056–1065. https://doi.org/10.1038/gim.2016.97

Jayashankar, S. S., Nasaruddin, M. L., Hassan, M. F., et al. (2023). Non-invasive prenatal testing (NIPT): Reliability, challenges, and future directions. *Diagnostics, 13*(15), 2570. https://doi.org/10.3390/diagnostics13152570

Johnston, M., Warton, C., Pertile, M. D., et al. (2023). Ethical issues associated with prenatal screening using non-invasive prenatal testing for sex chromosome aneuploidy. *Prenatal Diagnosis, 43*(2), 226–234. https://doi.org/10.1002/pd.6217

Kater-Kuipers, A., Bunnik, E. M., de Beaufort, I. D., & Galjaard, R. J. H. (2018). Limits to the scope of non-invasive prenatal testing (NIPT): An analysis of the international ethical framework for prenatal screening and an interview study with Dutch professionals. *BMC Pregnancy and Childbirth, 18*(1), 409. https://doi.org/10.1186/s12884-018-2050-4

Kater-Kuipers, A., de Beaufort, I. D., Galjaard, R. J. H., & Bunnik, E. M. (2020). Rethinking counselling in prenatal screening: An ethical analysis of informed consent in the context of non-invasive prenatal testing (NIPT). *Bioethics, 34*(7), 671–678. https://doi.org/10.1111/bioe.12760

Kotzee, B., Ignatowicz, A., & Thomas, H. (2017). Virtue in medical practice: An exploratory study. *HEC Forum, 29*(1), 1–19. https://doi.org/10.1007/s10730-016-9308-x

Labonté, V., Alsaid, D., Lang, B., & Meerpohl, J. J. (2019). Psychological and social consequences of non-invasive prenatal testing (NIPT): A scoping review. *BMC Pregnancy and Childbirth, 19*(1), 385. https://doi.org/10.1186/s12884-019-2518-x

Li, Y., Yang, X., Zhang, Y., Lou, H., Wu, M., Liu, F., Chang, W., & Zhao, X. (2024). The detection efficacy of noninvasive prenatal genetic testing (NIPT) for sex chromosome abnormalities and copy number variation and its differentiation in pregnant women of different ages. *Heliyon, 10*(2).

Lo, Y. M., Corbetta, N., Chamberlain, P. F., et al. (1997). Presence of fetal DNA in maternal plasma and serum. *The Lancet, 350*(9076), 485–487. https://doi.org/10.1016/S0140-6736(97)02174-0

Maftei, A., & Dănilă, O. (2022). The good, the bad, and the utilitarian: Attitudes towards genetic testing and implications for disability. *Current Psychology*, 1–22. https://doi.org/10.1007/s12144-021-02568-9

Mandal, J., Ponnambath, D. K., & Parija, S. C. (2016). Utilitarian and deontological ethics in medicine. *Tropical Parasitology, 6*(1), 5–7. https://doi.org/10.4103/2229-5070.175024

Mardy, A., & Wapner, R. J. (2016). Confined placental mosaicism and its impact on confirmation of NIPT results. *American Journal of Medical Genetics Part C: Seminars in Medical Genetics, 172*(2), 118–122. https://doi.org/10.1002/ajmg.c.31505

McGuire, A. L., Oliver, J. M., Slashinski, M. J., et al. (2011). To share or not to share: A randomized trial of consent for data sharing in genome research. *Genetics in Medicine, 13* (11), 948–955. https://doi.org/10.1097/GIM.0b013e3182227589

Middleton, A., Milne, R., Almarri, M. A., et al. (2020). Global public perceptions of genomic data sharing: What shapes the willingness to donate DNA and health data? *The American Journal of Human Genetics, 107*(4), 743–752. https://doi.org/10.1016/j.ajhg.2020.08.023

Minear, M. A., Alessi, S., Allyse, M., Michie, M., & Chandrasekharan, S. (2015). Noninvasive prenatal genetic testing: Current and emerging ethical, legal, and social issues. *Annual Review of Genomics and Human Genetics, 16*, 369–398. https://doi.org/10.1146/annurev-genom-090314-050000

Pertile, M. D. (2018). Genome-wide cell-free DNA-based prenatal testing for rare autosomal trisomies and subchromosomal abnormalities. In *Noninvasive prenatal testing (NIPT)* (pp. 97–123). Academic Press.

Petersen, A. K., Cheung, S. W., Smith, J. L., Bi, W., Ward, P. A., Peacock, S., Braxton, A., Van Den Veyver, I. B., & Breman, A. M. (2017). Positive predictive value estimates for cell-free noninvasive prenatal screening from data of a large referral genetic diagnostic laboratory. *American Journal of Obstetrics and Gynecology, 217*(6), 691.e1–691.e6.

Potter, B. K., Avard, D., Entwistle, V., Kennedy, C., Chakraborty, P., McGuire, M., & Wilson, B. J. (2008). Ethical, legal, and social issues in health technology assessment for prenatal/preconceptional and newborn screening: A workshop report. *Public Health Genomics, 12*(1), 4–10.

Society for Maternal-Fetal Medicine (SMFM), Rink, B. D., Dugoff, L., Kuller, J. A., & SMFM Publications Committee. (2025). Society for Maternal-Fetal Medicine Consult Series #74: Cell-free DNA screening for aneuploidies: Updated guidance. *Pregnancy, 1*(6), Article e70139. https://doi.org/10.1002/pmf2.70139

Royal College of Obstetricians and Gynaecologists. (n.d.). Non-invasive prenatal testing for chromosomal abnormality using maternal plasma DNA (Scientific Impact Paper No. 15). https://www.rcog.org.uk/guidance/browse-all-guidance/scientific-impact-papers/non-invasive-prenatal-testing-for-chromosomal-abnormality-using-maternal-plasma-dna-scientific-impact-paper-no-15/.

Sanderson, S. C., Brothers, K. B., Mercaldo, N. D., et al. (2017). Public attitudes toward consent and data sharing in biobank research: A large multi-site experimental survey in the US. *The American Journal of Human Genetics, 100*(3), 414–427. https://doi.org/10.1016/j.ajhg.2017.01.021

Skotko, B. G., Allyse, M. A., Bajaj, K., et al. (2019). Adherence of cell-free DNA noninvasive prenatal screens to ACMG recommendations. *Genetics in Medicine, 21*(10), 2285–2292. https://doi.org/10.1038/s41436-019-0485-2

Society for Maternal-Fetal Medicine (SMFM) Publications Committee. (2015). SMFM statement: Clarification of recommendations regarding cell-free DNA aneuploidy screening.

American Journal of Obstetrics and Gynecology, 213(6), 753–754. https://doi.org/10.1016/j.ajog.2015.09.077

Svirsky, R., Sharabi-Nov, A., Sagi, T., Meiri, H., Adi, O., Kugler, N., & Maymon, R. (2023). High sensitivity and specificity in fetal gender identification in the first trimester using ultrasound and noninvasive prenatal screening (NIPS) in twin pregnancies: A prospective study. *BMC Pregnancy and Childbirth, 23*(1), 812.

Tonkens, R. (2015). Parental virtue and prenatal genetic alteration research. *Journal of bioethical inquiry, 12*(4), 651–664. https://doi.org/10.1007/s11673-015-9650-8

van den Berg, M., Timmermans, D. R. M., Ten Kate, L. P., van Vugt, J. M. G., & van der Wal, G. (2005). Are pregnant women making informed choices about prenatal screening? *Genetics in Medicine, 7*(5), 332–338. https://doi.org/10.1097/01.gim.0000162876.65555.ab

van der Meij, K. R. M., Kooij, C., Bekker, M. N., Galjaard, R. J. H., Henneman, L., & Dutch NIPT Consortium. (2021). Non-invasive prenatal test uptake in socioeconomically disadvantaged neighborhoods. *Prenatal Diagnosis, 41*(11), 1395–1400. https://doi.org/10.1002/pd.6043

Vanstone, M., Cernat, A., Nisker, J., & Schwartz, L. (2018). Women's perspectives on the ethical implications of non-invasive prenatal testing: A qualitative analysis to inform health policy decisions. *BMC Medical Ethics, 19*(1), 27. https://doi.org/10.1186/s12910-018-0267-4

World Health Organization. (2024). *Guidance for human genome data collection, access, use and sharing*. World Health Organization.

Wu, X., Huang, H., Chen, X., et al. (2021). Impact of confined placental mosaicism on non-invasive prenatal testing and pregnancy outcomes. *Zhonghua Yi Xue Yi Chuan Xue Za Zhi, 38*(4), 335–338. https://doi.org/10.3760/cma.j.cn511374-20200803-00574

Zaami, S., Orrico, A., Signore, F., Cavaliere, A. F., Mazzi, M., & Marinelli, E. (2021). Ethical, legal and social issues (ELSI) associated with non-invasive prenatal testing: Reflections on the evolution of prenatal diagnosis and procreative choices. *Genes, 12*(2), 204. https://doi.org/10.3390/genes12020204

Index

'*Note:* Page numbers followed by "f" indicate figures and "t" indicate tables.'

D

E

O

P

Q

R

V

W

Printed and bound by CPI Group (UK) Ltd, Croydon, CR0 4YY
15/09/2026
02217074-0005